DEDICATION

This document i
Michelle Achieng Amani, Pendo Lorna
Adhiambo and Zawadi Faith Anyango

First published 2014
Set in Arial Bold

ISBN: 9966-7205-3-7

DECLARATION
All Photographs used in this book have been produced and processed by the author except where acknowledged.

PUBLISHED BY
WAMRA TECHNOPRISES,
P.O. BOX 36665-00200, CITY SQUARE, NAIROBI
TEL: +254722690956 / 738410345

1

CONTENTS

CHAPTER 1: INTRODUCTION

An item becomes a waste immediately it becomes worthless to the initial owner (Ali, 2003). However, somebody else may find this same material useful. Waste can thus be viewed as any material or resource whose next immediate use has not been identified. These may be solids, liquids or gases.

Definition: Solid waste management includes all activities that seek to minimize the health, environmental and aesthetic impacts of solid wastes (adapted from the SWM introductory text on www.sanicon.net (2003)

Solid waste is material, which is not in liquid form, and has no value to the person who is responsible for it. Although human or animal excreta often end up in the solid waste stream, generally the term solid waste does not include such materials. Synonyms to solid waste are terms such as "garbage", "trash", "refuse" and "rubbish".

The term municipal solid waste, refers to solid wastes from houses, streets and public places, shops, offices, and hospitals, which are very often the responsibility of municipal or other governmental authorities. Solid waste from industrial processes are generally not considered "municipal" however they need to be taken into account when dealing with solid waste as they often end up in the municipal solid waste stream.

Introduction
Human activities create waste, and it is the way these wastes are handled, stored, collected and

disposed of, which can pose risks to the environment and to public health. In urban areas, especially in the rapid urbanizing cities of the developing world, problems and issues of Municipal Solid Waste Management (MSWM) are of immediate importance. This has been acknowledged by most governments. However rapid population growth over-whelms the capacity of most municipal authorities to provide even the most basic services. Typically one to two thirds of the solid waste generated is not collected. As a result, the uncollected waste, which is often also mixed with human and animal excreta, is dumped indiscriminately in the streets and in drains, so contributing to flooding, breeding of insect and rodent vectors and the spread of diseases. Furthermore, even collected waste is often disposed of in uncontrolled dumpsites and/or burnt, polluting water resources and air. While urbanisation in developing countries has contributed to wealth accumulation, it has also been accompanied by an alarming growth in the incidence of poverty. Today, 25% of people in cities lives in absolute poverty, while another 25% is classified as "relatively poor". Throughout the developing world it is these urban poor, often in the peri-urban areas, that suffer most from the life-threatening conditions deriving from deficient MSWM.

Municipal authorities tend to allocate their limited financial resources to the richer areas of higher tax yields where citizens with more political pressure reside. Usually as income of the residents' in-creases, part of the wealth is used to avoid exposure to the environmental problems close to home, but as waste generation also increases with increasing wealth, the problems are simply shifted elsewhere. Thus even as environmental problems at the household or neighbourhood level may recede in higher income areas, city-wide and regional environmental degradation due to a deficient SWM remains or increases.

There are sometimes situations in which the difficulty experienced by urban managers in planning and directing concrete projects in a cost-effective way may overshadow the need for technical solutions to MSWM problems. In other cases, there is a tendency for MSWM decisions to be made without sufficient planning, to take into account only some aspects of a situation, to be based on a short-term view of the situation, or to be influenced by the interests of political elites. Adequate municipal solid waste management is much more than a technological issue. It always also involves institutional, social, legal, and financial aspects and involves coordinating and managing a large workforce and collaborating with many involved stakeholders as well as the general public. The preparation and management of a good solid waste management system needs inputs from a range of disciplines, and careful consideration of local conditions.

Challenges in Solid Waste Management
In Municipal Solid Waste Management (MSWM) of developing countries typical problem areas can be identified as: 1) inadequate service coverage and operational inefficiencies of services, 2) limited utilization of recycling activities, 3) inadequate landfill disposal, and 4) inadequate management of hazardous and healthcare waste.

Service Coverage for Waste Collection

Municipal solid waste collection schemes of cities in the developing world generally serve only a limited part of the urban population. The people remaining without waste collection services are usually the low-income population living in periurban areas. One of the main reasons is the lack of financial resources to cope with the increasing amount of generated waste produced by the rapid growing cities. Often inadequate fees charged and insufficient funds from a central municipal budget cannot finance adequate levels of service. However, not only financial problems affect the availability or sustainability of a waste collection service. Operational inefficiencies of SW services operated by municipalities can be due to inefficient institutional structures, inefficient organizational procedures, or deficient management capacity of the institutions involved as well as the use of inappropriate technologies.

With regard to the technical system, often the "conventional" collection approach, as developed and used in the industrialized countries, is applied in developing countries. The used vehicles are sophisticated, expensive and difficult to operate and maintain, thereby often inadequate for the conditions in developing countries. After a short time of operation usually only a small part of the vehicle fleet remains in operation. This complicates the collection system. This is worsened by lack of primary storage equipment, coupled by lack of awareness by the waste generators on their roles..

Private sector involvement in MSWM

In many countries there is currently great interest in involving private companies in solid waste management. Sometimes this is driven by the failures of municipal systems to provide adequate services, and sometimes by pressure from national governments and international agencies. Arrangements with private companies have not all been successful, and as a result some opposition to private sector involvement is now in evidence.

An important factor in the success of private sector participation is the ability of the client or grantor - usually a municipal administration - to write and enforce an effective contract. Many municipalities do not know what it has been costing them to provide a service, so they cannot judge if bids from the private sector are reasonable. The contract document must be well written to describe in quantitative terms what services are required and to specify penalties and other sanctions that will be applied in case of shortcomings. Monitoring and enforcement should be effective. It is also important that the rights of both parties are upheld by the courts. Three key components of successful arrangements are competition, transparency and accountability.

As an alternative to large (often international) companies that can provide most or all of the solid waste services in a city, microenterprises or small enterprises (MSEs) or Community-based Organisations (CBO) can be involved for services at the community level (neigbourhoods or the small city administrative zones). They often use simple equipment and labour-intensive methods, and therefore can collect waste in places where the conventional trucks of large companies cannot enter. The MSEs may be started as a business, to create income and employment, or they may be initiated by community members who wish to improve the immediate environment of their homes. A recurring problem with collection schemes that operate at the community level is that these systems generally collect and transport the waste a relatively short distance up to a transfer point, from where the waste should be collected by another organisation - often a municipality.

Problems of co-ordination and payment often result in the waste being left at transfer points for a long time creating a hygienic unsatisfactory condition. Another approach is to recycle as much of the waste locally (decentralised) so that there is very little need for on-going transport of collected waste. Resource Recovery and Recycling Recycling inorganic materials from municipal solid waste is often well developed by the activities of the informal sector although such activities are seldom recognised, supported, or promoted by the municipal authorities.

Some key factors that affect the potential for resource re-covery are the cost of the separated material, its purity, its quantity and its location. The costs of storage and transport are major factors that decide the economic potential for resource recovery. In many low-income countries, the fraction of material that is won for resource recovery is very high, because this work is done in a very labour-intensive way, and for very low incomes. In such situations the creation of employment is the main economic benefit of resource recovery. The situation in industrialised countries is very different, since resource recovery is undertaken by the formal sector, driven by law and a general public concern for the environment, and often at considerable expense.

Reuse of organic waste material, often contributing to more than 50% of the total waste amount, is still fairly limited but often has great recovery potential. It reduces costs of the disposal facilities, prolongs the sites life span, and also reduces the environmental impact of disposal sites as the organics are largely to blame for the polluting leachate and methane problems. This is one of the reasons why solid waste managers in many parts of the world are now exploring ways to reduce the flow of biodegradable materials to landfills. The feasibility of municipal solid waste composting as one step in the citywide solid waste management system depends on the market for the compost product, as well as the technical and organisational set-up of the individual plants. Last but not least, a clear legislation, policy and municipal strategy versus the management of organic waste is an important prerequisite for the success of composting activities.

Solid waste Disposal
Most of the municipal solid waste (MSW) in developing countries is dumped on land in a more or less uncontrolled manner. These dumps make very uneconomical use of the available space, allow free access to waste pickers, animals and flies and often produce unpleasant and hazardous smoke from slow-burning fires. Financial and institutional

constraints are the main reasons for inadequate disposal of waste especially were local governments are weak or underfinanced and rapid population growth con-tinues. Financing of safe disposal of solid waste poses a difficult problem as most people are willing to pay for the removal of the refuse from their immediate environment but then "out of sight – out of mind" are generally not concerned with its ultimate disposal.

The present disposal situation is expected to deteriorate even more as with rapid urbanization settlements and housing estates now increasingly encircle the existing dumps and the environmental degradation associated with these dumps directly affect the population. Waste disposal sites are therefore also subject to growing opposition and it is becoming increasingly difficult to find new sites which find public approval and which are located at a reasonable distance from the collection area. Siting landfills at greater distances to the central collection areas implies higher transfer costs as well as additional investments in the infrastructure of roads hence intensifying the financial problems of the responsible authorities. In addition to all this, an increase in service coverage will even aggravate the disposal problem if the amount of waste cannot be reduced by waste recovery.

Other reasons for inadequate disposal are the mostly inappropriate guidelines for siting, design and operation of new landfills as well as missing recommendations for possible upgrading options of existing open dumps. Many of the municipal officials think that uncontrolled waste disposal is the best that is possible. Often the only guidelines for landfills available are those from high-income countries. These are based on techno-logical standards and practices suited to the conditions and regulations of high-income countries and do not take into account for the different technical, economical, social and institutional aspects of developing countries. The safe alternative, a sanitary landfill, is a site where solid wastes are disposed at a carefully selected location constructed and maintained by means of engineering techniques that minimise pollution of air, water and soil, and other risks to man and animals. Loans or grants to construct sanitary landfills do not necessarily result in sanitary landfill disposal.

Equally important as site location and construction is well trained personnel and the provision of sufficient financial and physical resources to allow a reasonable standard of operation. If this is not given, good sites can quickly degenerate into open dumps. Healthcare Wastes from Hospitals and Hazardous Wastes Healthcare wastes are generated as a result of activities related to the practice of medicine and sales of pharmaceuticals. Some of the health-care wastes coming from any particular hospital or institution are similar in nature to domestic solid wastes, and may be called "general health-care wastes". The remaining wastes pose serious health hazards because of their physical, chemical or biological nature, and so are known as "hazardous healthcare wastes".

In many cases the most dangerous items in healthcare wastes are needles from syringes and drips, because the needles shield the viruses from chemical disinfectants and a harsh external environment, and the sharp point allows easy access for the viruses into the blood stream of anyone who is pricked by the needle. For treatment of hazardous healthcare wastes many strategies rely solely on the provision of incinerators or other treatment technologies. Such a strategy has several weaknesses as often the hospitals and healthcare facilities are not able to afford the operating costs of the plant. Thus plants are left unused or not repaired when they break down. Further, many of the risks occur before the waste gets to this final stage, and therefore they are not reduced by the provision of treatment equipment. The key to improving healthcare waste management is to provide better methods of storage and to train the staff to adopt safer working practices and segregate as hazardous healthcare wastes from general healthcare wastes.

Some waste materials need special care and treatment because their properties make them more hazardous or problematic than general wastes. The management of hazardous chemicals is not only a matter of technology and legislation, but also of enforcement, funding and financial instruments. Changing processes to use less hazardous substitutes and minimising hazardous waste quantities that are discarded can be seen as the preferred options in dealing with any difficult waste. The Basel Convention seeks to control the movement of hazardous wastes across international boundaries. This instrument is necessary because the high cost of treating hazardous wastes in industrialised countries makes it financially attractive to ship the wastes to another country where no special requirements for their disposal will be applied.

Factors Influencing Solid Waste Management in Developing Countries

There are many factors that vary from place to place and that must be considered in the design of a solid waste management system. These are discussed below.

Generation, handling, treatment and disposal
Brief Description of the Sector

Though high and low-value recyclables are typically recovered and reused, these make up only a small proportion of the total waste stream. The great majority of the waste (about 70 percent) is organic. In theory, this waste could be converted to compost or used to generate biogas, but in situations where rudimentary solid waste management systems barely function, it is difficult to promote innovation, even when it is potentially cost-effective to do so. In addition, hazardous and infectious materials are discarded along with general waste throughout the continent. This is an especially dangerous condition that complicates the waste management problem.

Throughout most of sub-Saharan Africa solid waste generation exceeds collection capacity. This is in part due to rapid urban population growth: while only 35 percent of the sub-Saharan population lives in urban areas, the urban population grew by 150 percent between 1970 and 1990. But the problem of growing demand is compounded by broken-down collection trucks and poor program management and design. In West African cities, as many as 70 percent of trucks are always out of service at any one time, and in 1999 the City of Harare failed to collect refuse

The adverse impacts of waste management are best addressed by establishing integrated programs where all types of waste and all facets of the waste management process are considered together. The long-term goal should be to develop an integrated waste management system and build the technical, financial, and administrative capacity to manage and sustain it.

Trends in Waste generation and the need for improved management strategies

In all communities, people generate all kinds of wastes at domestic, industrial, commercial, and farm levels. At the most basic level is domestic waste, which largely comprise food wastes, animal manure, ashes, broken tools and utensils, and old clothing. In an agricultural community, this waste is readily absorbed into the environment by the natural cycles. However, since the last century, there has been an increase in the number of people living in towns. Urbanization and industrial development have rapidly increased the range, diversity and quantity of wastes being produced. These require collection and disposal. To manage wastes properly, it is important to know: How much waste is being produced; the composition of waste from each of the different sources; and how the quantity and composition might change in the future. Wastes can come from domestic, commercial, institutional, street sweepings, construction / demolition and industrial sources.

CHAPTER 2: THE CURRENT CHALLENGES OF NAIROBI HOUSEHOLDS IN UNDERTAKING THEIR SWM TASKS

Abstract

In a research conducted in Nairobi in 2005/7 period, some very interesting findings came out. The study aimed at studying the challenges faced by Nairobi city households, with a view to building their capacity in improving Solid Waste Management (SWM). The objective of the study was to determine the challenges facing Nairobi households (HHs) in SWM. The approached used were qualitative and quantitative, comprising 12 key informant interviews, 6 Focus group discussions, transect walk, and 430 pre-tested household (HH) questionnaires were administered in proportionate to estate population among the high income, medium to high income, medium income and, low-income estates, each referring to different social status. All the 8 Nairobi divisions (now Districts) were represented, in a ratio proportionate to HH numbers as follows: (i) 0% of sampled HHs from high income estates; (ii) 90 (20.9 %) of sampled HHs from middle-high income estates of Langata and Riruta; (iii) 100 (23.2%) of sampled HHs from lower middle income estates of Makongeni and Kayole; (iv) 100 (23.2%) of sampled HHs from low income estates of Eastley, Kawangware and Komarock; and (v) 140 (32.6%) of sampled HHs from slum estates of Kibera, Korokocho, and Mathare.

The bulk of the challenges HHs faced in their efforts to improve SWM in Nairobi arose from what specific roles they actually played. Some of these activities included the emptying of the bins, storing, sorting, transporting and scavenging of wastes. Up to 64%, 53%, 50% and 46% of households had communal storage in slums, low income, lower middle income and middle-high income estates respectively. However, 78.4% of Nairobi HHs were not satisfied with the then state of SWM, as well as the service they got, while 30% din't get any SWM service at all. The key SWM challenges faced by households included lack of Government support (65.3%), Lack of cooperation among the neighbours (60.5%), lack of storage (57.2%), air-borne littering (32.3%), lack of Personal Protective equipment (PPEs) (29.3%) and long distance to disposal point (27.7%). The collection service varied greatly with socio-economic class of the estate, with the mean service providers for the entire Nairobi being Private firm (40%), nobody (30%) and Nairobi City council (NCC)(20%). All estates overwhelmingly concured that the state of existing dumpsites was poor. More than 75% of households in Kibera, Mathare, Kayole, Komarock, Kawangware and Korokocho rated the dumpsites as poor.

The majority of Nairobi HHs (39.5%) considered their full waste bin as of average weight, while 23% considered them as heavy. Scavenging was common in all estates regardless of class, with a mean of 82% HHs scavenging, of whom at least 70% of participants were children. This posed huge health and safety hazard, since at least 50% did not use any PPE, while Korokocho, Mathare and Kibera had more than 50% cases of non-protection. Up to 57.4% of Nairobi households stored their waste in thin use-and-dump plastic (Jwala) and 34.4% in bucket, yet 48.4% prefered a bucket bin and only 23.5% prefer jwala. This had partly worsened the plastic menace in the city. It was concluded that challenges faced by HHs varied from lack of Government support, Lack of cooperation among neighbours, health and safety, poor attitude towards SWM and lack of appropriate bins. It was recommended that the community be given education to enable them initiate a community based solid waste management (CBSWM), which can integrate all sectors, and as many stakeholders as possible. Further research should be done on the households' desired level of SWM service and effective demand for the same so as to make the service provision in the sector demand driven.

Introduction

Nairobi, the capital city of Kenya, was the area of study. It is the administrative and commercial capital city of Kenya, which is one of the East African Countries. It is located at the equator at 6000ft above sea level. It covers an area of 684 km^2, and is the smallest province in Kenya. Other Kenyan provinces are central, Rift Valley, Nyanza, Western, Eastern, Coast and North Eastern. Nairobi province is bounded by Rift valley to the West and South, Eastern province to the East, Central province to the North and North East (GoK, 2001 , 2003).

Nairobi is a varied city, with rapid urbanization amidst deteriorating economic, environmental and health conditions, with features and facilities of a modern city on one hand, and extreme pockets of poverty and destitution on the other hand (Ikonya, 1991; Gathuru, 1990; and GoK, 1985). For instance, it has Kibera, Mathare and Korokocho as major slums, among others, where about 2 million Nairobi residents lived yet occupying only 5% of the municipal residential land (JICA, 1998; GoK, 1994a, 1994b). Kibera then prided in being the largest slum in Kenya and sub-Saharan Africa, with more than 25% of the Nairobi population confined in only 250 hectares of land (GoK, 2003 and WSP, 2005).

Nairobi was then administratively divided into divisions or administrative units: Mathare, Westlands, Starehe, Dagoreti, Langata, Makadara, Kamkunji and Embakasi. By 1999 population and household census (GoK 2001), there were 3,079,000 people distributed in 649,426 households. These are distributed as follows in the Nairobi administrative divisions

There was a general disparity of incomes as well as population densities in Nairobi. The people living in the western suburbs are generally the more affluent while the lower and middle-income elements of society dominate the eastern suburbs. Nairobi displays a complex surface structure, making it difficult to decipher the distinct land uses of the city surface. Inevitably, there are wide variations in population density reflecting different land use patterns of six distinct land use divisions, namely; the Central Business District (CBD); Industrial Area; public and private open spaces; public land; residential areas; and undeveloped land. The spatially divided internal structure is based on land uses and income levels (Olima 2001, cited in Mitullah, 2003)).

Existing Literature on challenges of SWM in Nairobi.

Residential HHs are mainly interested in receiving effective and dependable waste collection service at a reasonably low price.

Disposal is not normally a priority demand of service users so long as the quality of their own living environment is not affected by dumpsites. In this respect, they tend to exhibit a 'Not In My Backyard' (NIMBY) syndrome (Afullo, 2004). Only informed and aware citizens get concerned with the broader objective of environmentally sound waste disposal.

Informal settlements is a term used interchangeably with slum communities (United Nations 1996, cited in Mary & Negussie (undated)). Informal waste workers usually live and work under extremely precarious conditions, including scavenging. Scavenging, in particular, requires very long working hours and is often associated with homelessness. Besides social marginalization, waste workers and their families are subject to social and economic insecurity, health hazards, lack of access to normal social services such as health care and schooling for children (Schübeler et al, 1996:23).

Low waste collection rate remain a challenge to all waste generators, most importantly, the low-income estate households. Collection systems comprise household and neighbourhood (primary) waste containers, primary and secondary collections vehicles and equipment, and the organisation and equipping of collection workers, including the provision of personal protective clothing (PPC) (Ali, 2003; Ali et al, 1996). Selection of collection equipment should be based on area-specific data on waste composition and volumes, local waste handling patterns and local costs for equipment procurement and O&M (Schübeler et al, 1996:48).

This poor state of SWM services was attributed to insufficient financial outlays, shortage of equipment and unfavorable institutional and organizational arrangements. There was also lack of systematic and integrated approach to tackling the waste management problem. The attitudes of poorer city residents toward environmental cleanliness was also a contributing factor (Kibwage, 1996, 2002). An urgent need existed for new methods of waste handling and promoting fuller environmental awareness among HHs. This should arise from knowledge of challenges they face, thus becoming demand driven.

Regarding the design of local waste collection systems, the most effective results may be obtained through the participation of the concerned communities. Where appropriate, the objectives of material recovery and source separation should be considered (Schübeler et al, 1996:48). Residential households want effective and dependable waste collection service at an affordable price. In low-income residential areas where most services were unsatisfactory, residents normally gave priority to water supply, electricity, roads, drains and sanitary services. Solid waste was (and still is, albeit less) commonly dumped onto nearby open sites, along main roads or railroad tracks, or into drains and waterways (Gathuru, 1993; City farmer, 2003; and Kibwage and Momanyi, 2003). Pressure to improve solid waste collection arises as other services become available and awareness mounts regarding the environmental and health impacts of poor waste collection service (Schübeler et al, 1996:21; Peters, 1998).

Households and community-based organisations (CBO) have important roles to play, not only as consumers or users of waste collection services, but also as providers and/or managers of local level services. In many low-income residential areas, CBSWM is the only feasible and affordable solution, thus requiring integrated and participatory approaches (Ruto, 1988; Mulaku and Siriba, undated).The introduction of community-based solutions calls for awareness building measures and organisational and technical support.

The design of MSWM systems must be adapted to the physical characteristics of the area. The interaction between waste handling procedures and public health conditions is influenced by climatic conditions and characteristics of local natural and ecological systems. The degree to which uncontrolled waste dumpsites pose fly, scavenger, aesthetic and rodent hazards depends largely on prevailing climatic and natural conditions. In general, climate determines the frequency with which waste collection points must be serviced to limit negative environmental consequences (Liyai, 1998; Komu et al, 1993; and Schübeler et al, 1996:27).

Key Challenges

The key SWM challenges faced by HHs included lack of Government support (65.3%), lack of cooperation from neighbours (60.5%), lack of storage (57.2%), air-borne littering (32.3%), lack of PPEs (29.3%) and long distance to disposal point (27.7%). Of these, storage and cooperation were key challenges in Mathare, Kawangware, Korokocho and Makongeni estates; Cooperation and government support are major challenges in Eastleigh, Kibera and Kayole while Government support was key challenge in Langata and Kibera estates. These challenges which affected the HH waste generators most needed to be looked into, with a view to creating a partnership and collaborating among all stakeholders in the SWM sector.

Who collects solid wastes from Nairobi estates

The extent of collection of garbage from Nairobi estates varied greatly with socio-economic class of the estate. The service providers for the entire Nairobi was Private firm (40%), nobody (30%) and Nairobi City council (20%). For most of the estates, private firm seemed to have taken over the responsibility. This applied to Eastleigh-55%, Makongeni-50%, Mathare-55%, Langata-89%, Kayole-38.3%, Komarock-95%, kangemi-52.5%, and Riruta-92%. Only Kibera and Korokocho, considered the worst of among the slums, had insignificant presence of private firm, with it no collection at all in 82% and 78% of HHs respectively.

Generally, the Nairobi city Council (NCC) then seemed to concentrate in Eastleigh (47.5%), Kangemi (37.5%), Kayole (26.7%) and Makongeni (25%) where it still was either second or last among significant waste collectors; its contribution did not exceed 40% households in all studied estates except in Eastleigh. JICA (1998) study indicated that 74% and 25% of high income and low income residents respectively were receiving SWM service. Given this is supposed to be NCC's key role, it can be considered a failure. It cannot be unfair to say NCC has collapsed.

All Eastleigh, Komarock and Riruta households had some collection going on, and it is significant that these were areas where there was heavy private firm presence. Langata, Kangemi and Makongeni estates had less than 16% of households with no collection service. However, Langata was already topping in private sector contribution (89%), while the other two estates had at least half of the households serviced by the private sector. It was clear from these findings that the future of SWM service in Nairobi estates lie in privatization. In terms of estates classification, 49.3% of households were served by private firm, 29.3% not served and only 19% by Nairobi city council. Among the slum estates studied, there was no collection service in 65% of households, while private form served 18.6% and NCC serves 11.4%. Among the low income estates, Private firm served 43%, NCC served 26% and 28% got no service. Among the lower-middle income estates, 62% were served by private firm, 35% by NCC and 3% got no service. For the middle-high income estates, 90% got private sector service, 5.6% were served by NCC, while 4.4% got no service.

About 60.5%, 34.4%, 30.2%, 13.3% and 7.4% of Nairobi HHs felt they had extra capacity in waste storage, transport, separation and sorting, reprocessing and scavenging respectively. This means that even though the HHs stated overwhelmingly as already playing key roles in storage (90.5%) and transporting (37.9%), they felt they were not doing as should be, and could improve. In roles which they were then only playing marginally such as separation (8.8%), reprocessing (3.3%) and scavenging

(1.4%), there was an increased willingness to play more role. For instance, scavenging improved from 1.4% to 7.4% (net 6% increase); reprocessing from 3.3% to 13.3% (net 10% increase), while separation increased from 8.8% to 30.2% (net 21.4% increase). This unconditional willingness by Nairobi HHs to increase their role in SWM was a gesture which could be tapped into and used to benefit all in the city and get rid of the rot.

Mathare and Kibera were slum estates where there was high level of scavenging and reprocessing already, thus accounting for the low response (0-4%). Willingness to sort was highest in Eastleigh (77.5%) and lowest in Mathare (2.5%).The latter was because the Mathare residents were already fully involved, with almost nothing more to sort, thanks to intensive scavenging from where majority eke a living.

Activities undertaken by Nairobi Households

Generally, the bulk of the challenges HHs faced in their efforts to improve SWM in Nairobi arose (and perhaps still) arises from what specific roles they actually play. In 2005/7 period, some of these activities included the emptying of the bins, storing, sorting, transporting and scavenging of wastes. Going by the socio-economic class, the HH may or may not have an official communal storage. 64%, 53%, 50% and 46% of HHs had communal storage in slums, low income, lower middle income and middle-high income estates respectively (figures 4, 5 and 6).

The majority (90.2%) of Nairobi HH waste bins were emptied by HH members, a figure fully contributed by Langata, Mathare, Kibera, Korokocho and Riruta at 100% HH. The other 9.8% was largely contributed by Komarock (40%), Eastleigh (27.5%), Kayole (18.7%), Kawangware (12.5%) and Makongeni (7.5%). In some cases, either HH or non-HH member emptied the waste bin. This makes the total % totals exceed 100%. The 9.8% of non-HH member participation represented a door-to-door collection, while the rest of the over 90% represented other collection methods.

Safety and security

All estates overwhelmingly concured that the state of existing dumpsites was poor, with the bulk of the residents considering the dumpsites as insecure, and poorly maintained. More than 75% of HHs in Kibera, Mathare, Kayole, Komarock, Kawangware and Korokocho rated the dumpsites as poor. Only 32.5%, 30%, 18.3%, 14%, 2% of residents in Makongeni, Eastleigh Kayole, Kibera and Riruta respectively considered the state of the existing dumpsite as fair. Less than 5% in all estates rated them as good, while none considered them very good to excellent. More than 66% of the HHs rated the dumpsites as of poor state; 10% considered them fair; 1.4% considered them good, and none rated them as very good. About 3.3% did not know the state.

All broad estate groupings rated the security of the existing dumpsites as poor, with slums, low income, lower middle and middle-high income estates registering 85%, 78%, 64% and 29% of their HHs rating security at the dumpsite as poor. All other responses were marginal and almost insignificant. In general, 66% of the entire Nairobi HHs considered the dumpsite as unsafe; 10% considered the security status as fair, while 2.1% considers them good. None considered the security status as very good or excellent.

Deficient Household containers

The majority of Nairobi HHs (39.5%) rated the weight of their waste bin as average. However, 23% considered them heavy, 22.6% light, 7.2% very heavy and 6.7% very light. These had serious implications on whether the wastes end up in the right place, and health and safety of those handling the full bins. Within Estates 65% in Eastleigh, 65% in Langata, 85% in Komarock, 52.5% in Makongeni described the weight of their full waste bin as average, while 50% in Mathare, and 90% in Korokocho describe their as light. The Mathare and Korokocho cases were slums where HHs got very little to eat, and therefore almost nothing to dispose of. The above is fairly satisfactory scenario since the burden of carrying heavy waste bins affects an insignificant minority in the above estates. However, 47.5% in Kawangware, 32% in Kibera and 37.8% in Kayole handled heavy waste bins. Of greatest concern were Kibera, Kawangware, Kayole, Eastleigh and Makongeni estates where over 30% of full bins were considered as heavy to very heavy by the HHs. In Mathare, Komarock, Korokocho and Riruta no more than 10% of HHs consider the full bins as heavy to very heavy. The significant group to take care of their needs urgently is the 30% whose full bin is either heavy or very heavy.

General Scavenging.

Scavenging then seemed to be common in all estates regardless of socio-economic class, with at least 70% of HHs being aware of scavenging activity. The average figure for Nairobi was 81.6% scavenging in their estate. Half (50%) of study estates registered above average figures of 90-100%, with all estate groups represented. The slums recorded the highest % of scavenging at 90%, then lower middle class and middle-high classes at 83% and 82.2 % respectively. This suggests that scavenging seems to be done across the board.

Household scavenging

Two thirds of the HHs do not scavenge while the wastes were still in the house, while 25.6% acknowledged having HH scavenging. A small% of HHs opted not to answer the question, partly due to the embarrassment the practice would cause to the respondents. In all estates, the no-scavenging response (No) exceeded the affirmative response (Yes), except in Kibera and Makongeni where 58% and 55% respectively of HHs agreed there was scavenging at HH level.

Nairobi Household scavengers

In Nairobi, 70% of HH scavenging was done by children, 20% by the maid and 10% by the private sector. In Eastleigh and Makongeni, only children scavenged in 5% and 30% of HHs respectively. In Kawangware, only the private sector did it marginally in 2.5% of the HHs; in Mathare and Riruta, only the children and maid scavenged at 10% and 16% of HHs respectively, while in Kibera, children scavenged in 50%of the HHs while maids did it in 12% of the HHs. This is part of a reason why there were no outside scavengers, indicated in an earlier section of this chapter. In Langata, Kayole and Komarock, all the three groups of scavengers were involved albeit to varying degrees. Lastly, in Korokocho the children and the private sector scavenged in 10% of the HHs.

Health and safety issues in Nairobi's SWM sector

Danger loomed (and it still does) for scavengers because most did not use any personal protective equipment (PPE), with Eastleigh and Riruta leading in non-protection. Other estates with more than 50% cases of protection were the slums such as Korokocho, Mathare and Kibera. This was partly attributed to the intensity of scavenging there, the duration it had been done, and increased level of awareness about risks among the scavengers and those they scavenged for. The latter would provide some rudimentary PPEs, their sufficiency notwithstanding. Scavengers interviewed indicated that putting on PPE gave them some confidence about their health and safety. Some scavengers also sold their PPEs, a practice which concured with past studies (GoK, 1994b; Bhutan, 2000; NEMA, 2005; Tenambergen, 1997; NCC, 2001; and ruto, 1998). Public sector waste workers and formal private sector workers were also subjected to unhealthy working conditions and poor social security as they participated in SWM. This remains a glaring gap, necessitating access to social and health care services.

Proper equipment and protective clothing can reduce health risks (Bhutan, 2000). By contributing to the "professionalisation" of the waste worker's role, proper clothing and equipment may also help to alleviate the social stigmatisation, which is often associated with waste work (Afullo, 2004; Tchobanoglous et al, 1993; UNEP, 2004, 2003, 1998; UNHCR, 1989; Thurguood, undated; and Schübeler et al, 1996:39). Informal sector waste workers are often socially marginalised and fragmented. They live and work without basic economic or social security, under conditions, which are extremely hazardous to health and detrimental to family social and educational development. Support to informal waste workers should aim to improve their working conditions and facilities, increase their earning capacity and ameliorate their social security, including access to housing, health and educational facilities. At the same time, the effectiveness of informal workers' contribution to the waste management may be significantly enhanced (Afullo, 2004).

Conclusions

The key SWM challenges faced by households varied (and still do) from lack of Government support, Lack of cooperation from among the neighbours, lack of storage bins. Four in five (78.4%) of Nairobi HHs were not satisfied with the then state of SWM, with 68% of the households considering the dumpsites as of poor state. Two in three (66%) of households considered the dumpsite as unsafe, with 30% of HHs not getting any SWM collection service in Nairobi. Close to three in five (57.4%) of Nairobi households stored their waste in thin plastic (Jwala), while 48.4% prefer a bucket bin and only 23.5% prefer jwala. This showed glaring gaps in terms of the households' desired level of SWM service, compared with what they were getting, if any.

CHAPTER 3: POTENTIAL CONCERNS AND ENVIRONMENTAL IMPACTS FROM SOLID WASTE MANAGEMENT ACTIVITIES

The typical municipal solid waste stream will contain general wastes (organics and recyclables), special wastes (household hazardous, medical, and industrial waste), and construction and demolition debris. Most adverse environmental impacts from solid waste management are rooted in inadequate or incomplete collection and recovery of recyclable or reusable wastes, as well as co-disposal of hazardous wastes. These impacts are also due to inappropriate siting, design, operation, or maintenance of dumps and landfills. Improper waste management activities can:

(i)Increase disease transmission or otherwise threaten public health. Rotting organic materials pose great public health risks, including, as mentioned above, serving as breeding grounds for disease vectors. Waste handlers and waste pickers are especially vulnerable and may also become vectors, contracting and transmitting diseases when human or animal excreta or medical wastes are in the waste stream. (See the discussion on medical wastes below and the separate section on "Healthcare Waste: Generation, Handling, Treatment, and Disposal" in this volume.) Risks of poisoning, cancer, birth defects, and other ailments are also high.

(ii) Contaminate ground and surface water. Municipal solid waste streams can bleed toxic materials and pathogenic organisms into the water bodies. (Leach ate is the liquid discharge of dumps and landfills; it is composed of rotted organic waste, liquid wastes, infiltrated rainwater and extracts of soluble material.) If the landfill is unlined, this runoff can contaminate ground or surface water, depending on the drainage system and the composition of the underlying soils. Many toxic materials, once placed in the general solid waste stream, can be treated or removed only with expensive advanced technologies. Currently, these are generally not feasible in Africa. Even after organic and biological elements are treated, the final product remains harmful.

(iii) Create greenhouse gas emissions and other air pollutants. When organic wastes are disposed of in deep dumps or landfills, they undergo anaerobic degradation and become significant sources of methane, a gas with 21 times the effect of carbon dioxide in trapping heat in the atmosphere. Garbage is often burned in residential areas and in landfills to reduce volume and uncover metals. Burning creates thick smoke that contains carbon monoxide, soot and nitrogen oxide, all of which are hazardous to human health and degrade urban air quality. Combustion of polyvinyl chlorides (PVCs) generates highly carcinogenic dioxins.

(iv) Damage ecosystems. When solid waste is dumped into rivers or streams it can alter aquatic habitats and harm native plants and animals. The high nutrient content in organic wastes can deplete dissolved oxygen in water bodies, denying oxygen to fish and other aquatic life form. Solids can cause sedimentation and change stream flow and bottom habitat. Siting dumps or landfills in sensitive ecosystems may destroy or significantly damage these valuable natural resources and the services they provide.

(v) Injure people and property. In locations where shantytowns or slums exist near open dumps or near badly designed or operated landfills, landslides or fires can destroy homes and injure or kill residents. The accumulation of waste along streets may present physical hazards, clog drains and cause localized flooding.

(vi) Discourages tourism and other business. The unpleasant odor and unattractive appearance of piles of uncollected solid waste along streets and in fields, forests and other natural areas, can discourage tourism and the establishment and/or maintenance of businesses.

DECISION MAKING GUIDELINES ON SOLID WASTE MANAGEMENT
a) How much waste? It is helpful to break this down into districts and different sources of waste e.g. household, industry, shops and institutions.
b) What is its general composition?
c) How might the amount or general composition of wastes change in the future?

CHAPTER 4: CLASSIFICATION OF SOLID WASTES

Domestic wastes:
These are produced from household activities, including food preparations, cleaning, fuel burning, garden wastes, old clothes and furniture, abandoned equipment, packaging and newsprint. Domestic wastes are also called household or residential wastes, and comprise a wide variety of materials. These include Food wastes, metal, plastic, glass, paper, rubber, textiles, ash, soil, pottery etc. From research, domestic food and ash dominate wastes in lower income countries, while in middle and high-income countries; there is a larger proportion of paper, plastics, metal, glass, and discarded manufactured items.

Commercial wastes
These are wastes from shops, offices, restaurants, hotels and similar commercial establishments. The wastes typically comprise packaging materials, office supplies, and food wastes, and have a close similarity to domestic waste. In lower income countries, food markets contribute to a large proportion of commercial wastes.

Institutional wastes
These are wastes from schools, hospitals, government offices and military bases. Their composition is close to domestic and commercial wastes although there is generally more proportion of paper than food waste. Hospital wastes inevitably include hazardous and infectious materials such as used bandages, sharp objects, like syringes and needles, and items contaminated with body

fluids. These may also be body parts and fluids. It is thus important to separate the hazardous and non-hazardous components in a health care waste to reduce the risk to health.

Street sweepings

Dust and soil, as well as varying amounts of paper, metal and other litter that is picked up from the streets dominate these. In some countries, street sweepings may also include drain cleanings, and varying amount of household and commercial waste dumped at the side of the road. These are also varying levels of plant remains and animal manure.

Construction and demolition waste

The composition depends on the type of building material used in a particular area, but is typically soil, stone, brick, wood, clay, reinforced concrete and ceramic materials. Some construction waste need disposal, but some may be recycled within the construction industry.

Sanitary waste (Night soil)

Where no sewage network exists within town, human wastes may be collected separately and used for agriculture or disposed of in landfills. If improperly disposed of, this material can contaminate water resources and be a source of infectious diseases.

Industrial waste

The composition of industrial waste depends on the industries involved. Much industrial waste is relatively similar to waste from commercial and domestic sources, and includes packaging, plastics, paper, and metal items. However, wastes from some industries are chemically hazardous. Disposal routes for hazardous wastes are usually different from those for non-hazardous wastes, and depend on the composition of each waste type. Hazardous wastes pose a risk of pollution drinking water supplies, watercourses or land or harming waste workers, and may need pre-treatment to reduce their toxicity if they are to be disposed of on land.

CHAPTER 5: WASTE CHARACTERISATION:
ASSESSING THE NATURE OF WASTE:
GENERAL TRENDS IN NATURE OF WASTE

In general, wastes generated per person in high income countries is much greater than in lower income countries. Typically, wastes as collected in high-income countries are less dense, as more packaging and lighter materials are discarded, and less ash and food waste. Moisture content is much higher in low-income countries' wastes because of the water in food wastes (which incidentally, also is the main cause of food going bad). Wastes from higher income countries with proportionately less food wastes have lower moisture levels.

The following characteristics of waste are commonly determined. The parameters are: waste density, composition, moisture content, size distribution generation rate and The reasons for their choice is also discussed under each item. An effective SWM plan requires full knowledge about the nature of the wastes. This can be determined by sampling. Solid wastes can then be characterized using the following criteria: Generation rate (ii) Moisture Content (iii)_ Density (iv) Composition (iv) Size Distribution

After assessing the quantity, the other characteristics of waste to enable one select the most appropriate waste disposal option include: Moisture content, biodegradability potential, heating value and density, composition, and size distribution. An effective Solid Waste Management (SWM) requires these details because they will affect choices concerning:

(i) Method of storage (ii) The method and frequency of collection; (iii) The equipment used for collection; (iv) The size of the workforce; and (v) The method of disposal.

In addition, the properties of the waste also indicate: The potential for resource recovery; and (ii) The environmental impact if the wastes are not properly managed.

While assessing the nature of wastes, one must consider not only the existing conditions but also the prediction of future trends. Also as the level of development of a country increases, the characteristics of waste produced changes. There are three general trends that are observable worldwide:

• Per capita waste generation rates increases with increase in economic development;

• The development of a country is closely followed by an increase in the amount of paper;

• The proportion of putrescible materials, ash and density decrease with development.

Waste generation Rate:

This aspect describes how quickly a certain quantity of waste is produced. . It is usually defined as the average amount of waste generated by one person in one day. In weight terms, the common unit is Kilograms per capita per day (Kg/cap/day) (hereafter called Kg/c/d).

Thousands of tons of solid waste are generated daily in Africa. Most of it ends up in open dumps and wetlands, contaminating surface and ground water and posing major health hazards. Generation rates, available only for select cities and regions, are approximately 0.5 kilograms per person per day—in some cases reaching as high as 0.8 kilograms per person per day. While this may seem modest compared to the1–2 kg per person per day generated in developed countries, most waste in Africa is not collected by municipal collection systems because of poor management, fiscal irresponsibility or malfeasance, equipment failure, or inadequate waste management budgets.

Waste composition:
The composition of the seven solid waste types is highly variable. They are influenced strongly by: climate and seasonal variation, the prevailing economy, the physical characteristics of the city, social and religious customs. Variations in solid waste composition are mostly significant when a municipality is making decisions about the suitability of a specific treatment or disposal method, such as composting or incineration. It is important for the waste manager in a city to have an indication of the composition of waste in the locality. Sampling of the wastes is technically possible, and must be done in one or more areas, which are statistically representative.

In this way, the information on waste composition and characteristics can be obtained to aid in decision-making about the viability of waste treatment processes, collection equipment changes and recycling initiatives. Flintoff, 1976, cited in Ali (2003), developed the following model of changes in waste quantities at different stages in the waste handling processes.

Table 1: Changes in composition and nature of solid waste along the stream (Source: Ali, 2003)

STAGE	HANDLING PHASE	LOSSES
TOTAL GENERATED		
1	Minus	Salvage sold by householder
2	Minus	Salvage by servants
3	Minus	Salvage by waste pickers in streets
= TOTAL COLLECTED		
	Minus	Salvage by collectors
=TOTAL DELIVERED FOR DISPOSAL		
	Minus	Salvage by disposal staff
	Minus	Salvage by waste pickers at disposal site
= TOTAL DISPOSED		

Representative sampling therefore requires an efficient SWM program. It requires an efficient system devoid of laziness and excuses. It requires a working system. These waste reduction processes determine the quantity and composition of deposited waste. They also determine the design guidelines for a new disposal facility, alongside any material recovery facilities.

Solid waste is a heterogeneous material, comprising many different components. It is the proportion of these different substances in the waste that composition analysis describes. The composition is described as the percentage of each component present in the waste. These components may include paper, plastic, wood, glass, metal etc. it is calculated based on wet weight proportion, and thus should be done before the wastes dry up, or are exposed to any changes. Ali (2003) attributes changes in composition of waste to one or a combination of the following:

- Natural biodegradation and volatilization of waste constituents;
- Picking out of recyclables by human scavengers;
- Eating of food wastes by animal scavengers;
- The burning of waste;
- Rainfall soaking and / or leaching the wastes (personal observation);
- Animal manure being added as the animals scavenge;
- Using storage sites as toilets and adding night soil; and
- Waste generating sectors such as industries throwing waste in with the rest.
- Adding of soil and stone from street sweepings and construction / demolition.
- Addition of stabilized or semi-stabilized sewage and / or industrial sludge;
- Waste food eating by street children.

The waste composition, and more so moisture content affects (i) Biodegradability potential and (ii) calorific value, among others.

Biodegradability potential:
This is the proportion of biodegradable material in the waste. It forms single most important aspect of the composition of waste. It can be calculated by first removing non-biodegradable organic waste like plastic and rubber, then drying the waste at a temperature high enough to burn off the remaining organics.

Calorific value:
Is the amount of heat energy that can be produced if all the combustible parts of the waste are burned. If a town or city is considering waste incineration, this information

is crucial to establish whether the waste will burn without the use of additional fuel such as oil or gas.

Variations in Waste characteristics

Typically, domestic waste from industrialised countries has a high content of packaging made of paper, plastic, glass and metal, and so the waste has a low density. In many developing countries wastes contain large amounts of inerts such as sand, ash, dust and stones and high moisture levels because of the high usage of fresh fruit and vegetables. These factors make the waste very dense (high weight per unit volume). The consequences of this high density are that vehicles and systems that operate well with low-density wastes in industrialised countries are not suitable or reliable when the wastes are heavy. The combination of the extra weight, the abrasiveness of the sand and the corrosiveness caused by the water content, can cause very rapid deterioration of equipment. If the waste contains a high proportion of moisture, or is mostly inert material, it is not suitable for incineration, and so this treatment option is ruled out. Recycling or salvaging operations often reduce the proportion of combustible paper and plastic in waste before it reaches the treatment stage.

Access to waste for collection:

Many sources of waste might only be reached by roads or alleys which may be inaccessible to certain methods of transport because of their width, slope, congestion or surface. This is especially critical in unplanned settlements such as slums or low-income areas and thus largely affects the selection of equipment. Awareness and attitudes Public awareness and attitudes to waste can affect the whole solid waste management sys-tem.

All steps in solid waste management starting from household waste storage, to waste segregation, recycling, collection frequency, the amount of littering, the willingness to pay for waste management services, the opposition to the siting of waste treatment and disposal facilities, all depend on public awareness and participation. Thus this is also a crucial issue that determines the success or failure of a solid waste management system. Institutions and legislation Institutional issues include the current and in-tended legislation and the extent to which it is enforced. Standards and restrictions may limit the technology options that can be considered. The policy of government regarding the role of the private sector (formal and informal) should also be taken into account. The strength and concerns of trade unions can also have an important influence on what can be done.

Moisture Content
This is the percentage by weight of waste that is water. Drying a known amount of waste and measuring the weight change can determine this. Moisture Content (MC) is a measure of the amount of water in the waste. It is usually expressed as a percentage of the weight of water in a substance compared to its total weight. It is measured by noting the loss of weight after drying. MC depend on the presence of certain components, and is often related to the proportion of food wastes in a sample, as this component often has approximately 70% MC. MC is a vital factor in the choice of final disposal of waste. For instance, wastes with

high MC cannot be incinerated; composting requires a certain amount of moisture; while sanitary land filling depends on bio-degradability of waste, which is affected by MC. The level of moisture in a waste also determines the generation of polluted water (Leachate) and breeding of flies.

Waste density
The density of an object is the mass of an object made of that material divided by its volume. It is normally measured in Kilograms per m^3 (Kg/m^3). For solid waste, the term bulk density should be used since there is a considerable amount of air between the individual pieces of waste material. Bulk density includes the volume of this interstitial air. The bulk density of waste increases as the waste is compresses, or if water is added to it (Ali, 2003). Density of waste can be used in conjunction with generation rates to estimate the volume generated. This can vary from 0.4-1.0 liters / person/day (lcd). The volume and density of waste are important in choosing both the size and type of containers and collection vehicles. For instance, if a container of a certain capacity is chosen, the density can be used to calculate its weight when it is full to indicate whether it will be possible for one man to lift the full container.

The average density of solid wastes tends to decrease with level of development. In developing countries, there are lots of bulky waste material such as organic matter, mud, ash and soil. Besides, there is a lot of scavenging as some material is salvaged for sale, reuse or recycling. The ash volume increases because of widespread use of solid fuel. This is important in each stage in collection and disposal. For planning waste collection, the density achieved in each collection vehicle influences how many are needed to collect the waste in a particular area. Waste density is also important in a landfill site as an indication of how much space each delivery of waste will take before and after compaction and as a consequence, how long the total landfill space will take.

Size Distribution

Waste comprises many separate objects of different sizes. The size may influence the collection and disposal methods used, such as the diameter of storage bins. Large wastes may sometimes require size reduction by breaking up through shredding. This may facilitate compressing at the landfill, decomposition during composting and / or packaging in collection and storage bins. This may lead to a more uniform waste that may be easier to handle and process. Wastes with fine constituents such as sand may require screening prior to any further processing, as sand and ash can cause serious abrasion if the waste is made to slide against parts of a vehicle body. This may lead to quicker depreciation of the vehicle.

For health reasons, waste in tropical regions should actually be collected daily. This makes the challenges and costs of solid waste management in much of Africa even more daunting. It is generally the city center and the wealthier neighborhoods that receive service when it is available. In poorer areas, uncollected wastes accumulate at roadsides, are burned by residents, or are disposed of in illegal dumps that blight neighborhoods and harm public health. Where present, Manual Street sweeping by municipal employees or shopkeepers may help reduce these effects in the most public areas. Nonetheless, roadside accumulation in many cities has reached levels resembling those that spawned epidemics in European cities 500 years ago. Unless more effective urban waste management programs and public water supply systems are put in place, outbreaks of cholera, typhoid and plague may become increasingly common.

Only a small amount of the region's waste is disposed of in sanitary landfills; most is deposited in open dumps or semi-controlled unlined landfills with no groundwater protection, leachate recovery, or treatment systems. The larger dumps are located on the edges of cities, towns, and villages, sometimes in ecologically sensitive areas, or areas where groundwater supplies are threatened. They serve as breeding grounds for rats, flies, birds and other organisms that serve as disease vectors. Smoke from burning refuse may be damaging to the health of nearby residents and the smell degrades their quality of life. While the recovery and reuse of materials is generally for personal use, there are also many professional waste pickers. They are seriously threatened by disease organisms, sharp objects and other hazards in the waste, especially since they generally lack protective equipment. The high level of reuse of non-organic waste reflects the extent of poverty in the region.

Separation and treatment of organic waste is very rare. Municipal composting programs exist in some South African cities, but the few large-scale composting facilities built elsewhere are no longer operating. Anaerobic digestion to produce methane is not widely applied, and then usually uses manure, not organic waste. While solid waste collection is generally a municipal function, some countries and municipalities are now experimenting with limited privatization of these services, with some success. Because of the poor levels of collection, many residents—from impoverished to wealthy—pay for private collection of their wastes where these services are legalized. Municipal waste incinerators are too expensive for most communities and are not used. In any case, they are generally not practical, since most paper that can be reused from the waste stream is removed, leaving behind an organic waste that is too wet to burn. Some hospitals and municipalities have incinerators for medical waste, but these are often not operated correctly. The HIV/AIDs epidemic has raised concerns about reuse of syringes, and efforts are being made to construct low-cost, high-temperature two-chamber incinerators to destroy syringes along with other medical wastes.

Sector Design—Some Specific Guidance
Experience and study of solid waste collection programs in various parts of the developing world have identified a set of program elements and common pitfalls as well as a number of operations strategies to meet operational requirements and avoid commonn problems. Successful program:
• Apply an integrated holistic approach that takes into account key factors affecting waste generation, storage, and final disposition;
• Securing or establish stable financing and ensure funds are used appropriately;

• Carefully design, develop and implement privatization schemes after weighing the potential costs and benefits;
• Involve the community in waste-management decision making; and
• Build capacity of administrative and technical staff in government, NGOs and/or the private sector.

CHAPTER 6: INTEGRATED SOLID WASTE MANAGEMENT (ISWM):

This section describes a stage-by-stage analysis of the waste management chain. These stages include primary storage (at the source), primary collection, communal storage, secondary collection and disposal. Also discussed are some important aspects as waste handling at different stages, including the possible advantages of incorporating a transfer station, regardless of size and category.

Primary storage
This is the storage at source, e.g. at household. The most common primary storage containers are thin plastic bags, 5 or 10 litre plastic buckets, enclosed metallic or plastic 50-100 litre containers, etc. A few business premises use enclosures.

Primary Collection
A refuse collection service requires vehicles and labour. However, at this primary stage, vehicles are only involved when door-to-door, block or kerbside collection methods are employed. Primary collection involves taking / handling, carrying and depositing of waste from yards / households to the communal storage facilities. The method varies with size of container, distance from communal storage facility, pre-arranged collection method, and nature of premises etc. Commercial premises normally take their own wastes directly to the dumping ground using either a private company, or their own pick-up vehicles. A limited amount of waste in government, commercial and other private properties are collected by trucks.

Collection method

Typically, waste collection involves three main aspects. These are:

(i)   Travel to and from work area;

(ii)   The collection process, which includes: transfer of the wastes from storage to the vehicles and travel between successive collection points; and

(iii)   The delivery process whereby the contents of the vehicle are transported to the disposal site.

Travel to and from work area

Generally, the workers live in their own private quarters where they operate from daily. They would normally assemble at the office premises, while the driver gets the truck. The parking point also acts as the overnight packing area.

The crew, normally including the driver and one to two other members, would drive to communal storage facilities where they would find most of the waste already packed properly in the demountables. The waste is placed in these demountables by the public as well as by other support staff who would collect all scattered waste and place it in black film plastic containers, and placed in the demountable. The truck has a means of loading the demountable without the need for the workers to do the work by hand. This saves time.

Door to door collection

This is whereby the crew collects the waste from the house yards into awaiting vehicles outside the yard. It is rare, and is mainly practised in institutions where houses are close to one another.

Block collection

This is whereby the waste is taken from premises by the generators to an awaiting vehicle outside yards. This needs proper timing, as it cannot work if most households have nobody to deliver the wastes.

Kerbside collection

This is whereby wastes are placed in some container or site at the roadside, or at a kerb, awaiting collection. It is common in the roadsides e.g. at bus stops where alighting passengers, pedestrians and travellers waiting for vehicles would deposit their wastes.

Communal collection method:

This involves the placing of wastes at an official communal storage site awaiting collection by the council workers. It is the most common in residential sites, and a few commercial sites. It always goes along with a communal storage facility.

Communal storage

The Commonly used facilities for communal storage include Depots, enclosures, potable plastic bins and 200 litre drums. However, they are open, and animal (especially donkey) scavenging is common, accompanied with scattering of waste. This endangers the lives of the consumers of meat from these animals, as well as posing risks of passing zoonoses. This can be partly solved by public education, as the public due to negligence, and / or improper design, and location leaves these facilities open. The design can be improved to make it easier to close the doors, especially of enclosures and depots. Urination (and sometimes defecation) is also rampant at these facilities, especially enclosures and depots.

There is need for public education, coupled with provision of affordable public toilet facilities. They should be privatized to cater for 24-hour public toilet service.

Secondary Collection

The vehicles taking the wastes to the final dumping ground do this. In the private companies, the waste is collected mostly with bare hands, with no Personal Protective Equipment. The crew would then stand on the uncovered waste at the back of the vehicle. They would be exposed to all the smell and fly / vector / larva hazards, aside from the airborne littering which occur as the smelly waste-loaded vehicles travel at high speed. Some of this litter accumulates in sewers and storm drains. There is need for training of the SWM crew, provision of appropriate PPE, and regular clearing of sewer lines and drains to avoid blockage. Also, covered vehicles could reduce the problem of airborne littering.

Collection method

To be in good condition, and properly managed, or a combination of these three. Collection is irregular, leading to waste accumulation in many communal storage points. These heaps have blocked some paths and surface drains; interfered with drainage- causing flooding during rainy season; caused fly, aesthetic and odor menace; posed health risk to SWM crew, among other problems. There is need for the council to do preventive maintenance so that the SWM vehicles are in good condition, coupled with closer supervision, training and motivation of the crew. The work should be privatized. There is a lot of idle manpower, especially school-leavers who can do the job very well if more private companies were allowed into the sector. Some commercial and learning institutions collect and take their wastes to the dumping ground using either private companies, or their own vehicles. This initiative and effort should be encouraged, and the government, the council and the public should support the volunteers.

Vehicles for collection of garbage
Transport system for SWM
Vehicles to use in collection
Primary collection would involve the use of non motorized vehicles because Kadimo is still too small to have wide enough passages for alternative means. However, these may deliver separated wastes to a transfer station/ communal collection point from where motorized vehicles would come for them. The non-motorised means considered here are the hand carts and animal carts.

Animal drawn carts:
These would be useful in rough, sloppy areas. They are also larger than handcarts; they demand no fuel; can go longer distance compared with hand carts (3-4 km); are low cost compared with motor vehicles, produce no noise and the driver can leave the vehicle and assist in loading. However, their use would be less appropriate because (i) slow speed (ii) interfere with traffic (iii) animals have to be taken care of to work, e.g. fed, treated etc. They should take load to a two-level transfer station   at which they tip their loads directly into a motorized vehicle at a lower level. This, however, beats the finer and more desirable purpose of a transfer station- picking, sorting etc. Secondly, the handcart would be a new

technology in the area; animal training needs skills and it may be difficult in an area where residents are not used to it. The animal power will displace the cheap human labour, thereby contradicting the poverty reduction goal of the project; one animal can easily displace two men from work. Thirdly, an abattoir would create an automatic demand for animal products, thereby rendering animal power less attractive. Thus a hand cart is more recommended in this scenario.

Handcarts:
Condition and standard of use:
The minimum population density for which hand carts could be used is 7200 / km^2, when each km^2 would require 6 collectors and one trailer / transfer point. Surge capacity would not be necessary and thus the ratio of trailers to tractors would be 7:1, 6 trailers at transfer points and one being towed at any given time by one tractor. The use of short range transfer based on handcarts is relevant with: low per capita generation of wastes; high waste densities; high population density and low wage rates (Flintoff, 1976 pp 67).

Case study: Scenario in the entire Kadimo area:
Area = 232 km^2; current population = 12,920; projected population in 20 years = 21,000; current population density = 55.86 (56/ km^2); projected population density =90.5 (91/ km^2). Therefore the use of a handcart may not apply to the entire Kadimo area. It therefore must be analysed by sections e.g. town, major villages and scattered villages to see where handcarts would be relevant and where not.

Scenario in Kadimo town:

Area = 0.715 km^2; current population =4380; projected population in 20 years =9498; current population density = 6126/ km^2 projected population density = 13283/ km^2). Therefore the use of a handcart may not apply to Kadimo town now. Interpolation shows that the 7200 people / km^2 would be reached in the 3rd year (i.e. 1074 x 20 / 7157). Therefore since year 3 is still a preliminary year, it would be important to incorporate handcarts in the design of Kadimo town. The rural villages would not achieve the population density of 7200 / km^2 even in the 20th year. So a village solid waste management designs will not incorporate hand carts.

Therefore in Kadimo town, handcarts will form part of the design. They are more recommended because (i) the technology already exists among residents; (ii) it is cheap; (iii) Can pass through small passages which exist in Kadimo, and in the villages; (iv) Cause minimum obstruction (v) their capacity is enough to keep sweeper busy for up to two hours (vi) that are convenient in house to house collection especially in very narrow streets inaccessible by motor vehicles. However, they only carry small weights of up to 200kg, and cover small distances.

If 1 km^2 would require 6 collectors and 1 trailer/ transfer point, then in the entire Kadimo town (0.72 km^2), the following will be required: 6 waste collectors; 1 trailer / transfer point; Surge capacity would not be necessary and thus the ratio of trailers to tractors would be 7:1, 6 trailers at transfer points and one being towed. The use of short range transfer based on handcarts is relevant in Kadimo because of low per capita generation of wastes (0.4 kg/capita/day); high waste densities (300 kg/m³); high population density (>6000 in year 1, reaching 7200 in year 3, and > 13,000 in 20th year) and low wage rates (50,000-60,000 Ad/month) (Flintoff, 1976 pp 67; WEDC, 2005). Therefore a transfer station applies in Kadimo town.

The exchange bin system for separated wastes:
The hand carts will collect separated wastes from houses and kerbs, at which points the wastes will already have been separated and in smaller colour coded plastic containers as shown above. These will be emptied into the larger colour-coded cylindrical bins in the carts, according to their type. To save time and labour, the loads in the cart should be placed in a 4-6 colour-coded discrete bins which can be lifted off the cart and emptied or placed directly into a motorized vehicle at a transfer station preferable with a depot which contains replacement bins. Each colour would be for a specific type of waste, e.g. light plastics, heavy plastics, glass, food remains, metals, and mixed wastes. Cylindrical shape is emphasized here to save space in the cart. This can take the light tubular steel framework with a platform on which 3-6 bins of about 60 litres can be carried. The hand carts take the wastes within 3 hours to a communal storage (CS), where they leave the full 60-litre containers and pick fresh ones for use the following day (The exchange bin system) . The communal storage points are placed 2 km apart, so that the cart at the furthest end would be pushed for 1 km. The communal storage points also act as short range transfer stations. All cart operators should reach the site by 12 pm, so that the motorized vehicle takes these to a transfer station by 1 pm. This still ensures the wastes

are still in good condition for handling at the transfer station. The motorized vehicle would carry empty 60-litre bins for exchange with the full ones brought by the carts. The cart operators would provide the labour of carrying the bins to the motorized vehicle. This requires that a work study is done before each area is planned for.

Six-bin handcart:
Here, generation rate is likely to be low since residents recycle and reuse most of the wastes. Thus a six bin handcart can be used in place of a handcart or animal cart. It is convenient where generation rate is low, has a capacity of 300-500 litres of waste, and bins can be transferred directly to a transfer station- into a waiting vehicle (trailer-tractor) to take to a dumping point (but can still be taken to communal storage points as in the case of town).

Health and safety:
There should be enough good quality PPE for each crew member. They should be trained on the necessity of using them, and close supervision provided to ensure PPEs are properly used. Any failure would be warned, and sacked if need be. A supervisor with a motorcycle should be assigned to each region to do spot checks on collection cart crew. In Town, house to house or kerbside collection by handcarts-transport to communal storage within 1 km radius Daily collection would be recommended to avoid food remains rotting and becoming more difficult to handle.

Labour estimates:

In Kadimo, design population = 21,000. If each household has 6.5 persons, the number of households = 3231 households or dwellings. Half of these are in town, and half in periurban. Assume that in 20 years, the scattered villages will be the main settling areas as more residents look for land to establish private dwellings. Kadimo is a rural town. Therefore the villages are even more rural. This means the distance between houses may be a little far. In a typical house to house collection system, one 6-bin handcart load would cover 50 dwellings daily. In this scenario, we will assume half the rates, with one handcart covering 30 dwellings (HHs) daily. At a density of 300 kg/m^3, the weight per load would be about 120 kg, excluding the cart. Two people would use this well because Kadimo is slopy, though the cart is enough for one man. The 2 people per cart can be afforded because labour is cheap, and it takes care of emergencies. The load can only be managed by a cart with good design of bearings, which can be ensured locally.

With Kadimo town population of 9598, and an estimate of 0.4 kg waste / capita per day, the total solid waste produced equals 3,840 kg/ day = 3.84 tons/day @ 300 kg.m-3 gives 12.8 m3/day of waste. If I cart carries 120 kg load / day, a total of 32 carts would be required per day to take care of the solid wastes in the entire Kadimo town. One cart can be managed by one person. However, the terrain of Kadimo is rugged, and may require pushing the cart in some instances. For design purposes, two people are assigned a cart per day. This means that 64 workers (unskilled labourers) would handle the carts. At the rate of 50,180 Kshs / month/ labourer, 64 labourers would earn

3,211,520 Kshs per month (38,538,240 Kshs per year by the 20th year.

Personal protective equipment (PPE) costs: Assuming full PPE per worker costs 2,000 Kshs, up to 128,000 Kshs would be needed for full PPE for cart workers alone.

Cost of training staff on health and safety: This will be necessary every 6 months. 1-day in-training of all staff by private sector professional: 400,000 Kshs (monthly salary of a senior engineer) x 2 = 800,000 Kshs / year. Developing a health and safety policy with support of same private firm costed at 600,000 Kshs.

Type of handcart:
Box type handcart is recommended as preferred option because (i) it is cheaper, each costing 0.11 million Kshs, (ii) with less forex demand of 10%. This is compared to a handcart with 3 removable bins, which costs 0.28 million Kshs, with a forex component of 30%. Thus 32 simple box type handcarts would cost 3.52 million Kshs, with 0.352 million Kshs as forex component, while 32 handcarts with three removable bins would cost 8.96 million Kshs, with 2.688 million foreign exchange component. The latter is therefore clearly unnecessarily expensive. Secondly, despite the cost-effectiveness of a box handcart, it has a higher capacity of 0.8 m3, while the more expensive handcart with three removable bins has a capacity of only 0.27 m^3. Thus even the 32 such carts would not be enough for the task. Every one box type cart would have a capacity of 3 hand carts with three removable bins. The former can therefore carry up to 9 bins, or

between 3 and 9 larger bins. With the proposal of having about 4 colour coded cylindrical bins, the simple box handcart would serve the role best.

UN Habitat estimates that 100-300 m^3 of transport capacity would be required per million people in a developing country (UNHCR pp 15). If we assume a maximum figure of 300 m^3 for safety, (even though we know developing countries produce very dense wastes) 9598 people in Kadimo town would require 9598 x 300 m3 / 106, = 2.88 m^3 . With a box type handcart of capacity 0.8 m^3 each, 4 such carts would be enough to serve Kadimo town. This would only cost 0.44 million Kshs, with a 0.044 million Kshs as forex component. On the other hand, 12 handcarts with 3 removable bins would be required, costing 3.36 million Kshs, with a forex component of 1.008 million Kshs. This same amount would buy more than over 30 simple box type hand carts. Therefore clearly, the simple box type handcart should be purchased.

Since analysis from weight and volume perspectives give different numbers of carts, It is better to look at it from the weight perspective than the volume perspective, as the former brings in an element of efficiency. The cart with only 3 removable bins also limits separation, as there is likely to be more than three types of wastes. The box type of cart would, on the other hand, be more flexible, and offer thrice as much space, thereby able to accommodate more colour coded containers for a wider variety of separated wastes. Therefore initially, 8 number box handcarts should be bought immediately, but this should have reached 32 by the 20th year.

Handcart design: These can be simple type or a modified type.
Simple handcart:
A simple box cart with no partitions, and where all wastes are simply heaped. This is cheap and easy to construct and operate, but does not facilitate waste separation. It also makes it more difficult at transfer station, as more handling is involved, increasing the hazards. It is therefore discouraged.

Modified handcart:
This avoids the need to empty the cart on to the ground at transfer station, thus reducing the task of shoveling the wastes into another vehicle. The exchange bin system can be adopted to make work easier. The handcart can be equipped with portable receptacles that can be lifted off and emptied by one man into a transfer station serving a number of handcarts. The handcart can have:
(i)   Frame of light tubular steel, or angle, supporting a platform on which replaced two or more potable bins (6 preferred);
(ii)   wheels of a larger diameter, with rubber tyres, preferably pneumatic, ball or roller bearings;
(iii)   Portable bins of capacity 50-80 litres each, (60 preferred because wastes are dense);
(iv)   Brackets should be mounted on the frame of the handcart to carry three brooms and a shovel for sweepers;
This type of cart can be made locally, helping build capacity and saving foreign exchange.

Bins for the hand carts:

Assuming that one bin carries six 60 litre plastic bins with lid, another 6 would be required for daily exchange. Thus twelve 60 litre plastic bins with lid are required per handcart. This gives 384 bins in the 20th year. With each 60 litre bin costing 4700 Kshs, a total of 1,804,800 Kshs is required to buy bins with lid by the 20th year, which would comprise 180,480 Kshs as foreign exchange component. The table below outlines some difficulties likely to be met in the collection process, and how they should be resolved. Below are some constraints likely to be met at the collection stage, and ways of managing them.

Given the low population density in the village, it is not economical to have handcarts, or vehicles. Therefore simple baskets will be used to take wastes to a central receptacle (demountable) at a communal storage point, from where it will be picked by a trailer-tractor every 2 weeks. The wastes from the village will be less, high density and highly recycled. They can be taken and disposed of directly to the landfill.

Secondary / Communal storage:

This option looks at various aspects of communal storage (CS). Communal storage is a common facility where waste generators within a community take their wastes awaiting collection by other means for subsequent disposal or to a transfer station. It serves as a convenient short range transfer station where, in this case study, hand cart crews would deposit daily wastes by noon. One option is not to have any structure, since the wastes are brought in potable containers which are then loaded directly into a motorized vehicle bound for the transfer station in an exchange bin system. This is convenient because it is cheap. However, the waiting point should have a simple structure to act as shelter in case the crew is caught up by rain while waiting / loading the wastes. Thus the realistic options looked at here include: no structure; depots; and demountables.

Depots would be convenient, as they can be made from local materials. However, they can act as sites for urination, defecating, thuggery and even rape. They would also be associated with smell and fly hazards. Demountables would be heavy, and need strong motorized vehicles to mount, rendering it expensive. It also does not offer any chance for shelter, thus necessitating a second structure alongside it. It is a good alternative in villages where (i) people generate wastes very slowly- thus can be emptied directly to the disposal site every 2 -4 weeks by a tractor / trailer. It receives mixed wastes, thereby making it difficult and unhygienic for the crew to handle, especially if it is open for rains to fall into. This would catalyse rotting of decomposable matter, which is bound to be in large amounts in town. Therefore in town, only a simple shade would do. However, some constraints are expected whichever the secondary storage method adopted. These are as given in the table below.

The design of handcarts and animal-drawn carts

This may be complicated because of the slow speed, low capacity of the carts, and cultural barriers to the use of some types of animals in pulling carts (or performing some tasks).

However, carts are versatile and can perform any errand- from solid waste, to water transport, to agriculture produce transport, to ploughing (i.e. an animal trained in pulling of carts will need minimal training, if any, in ploughing). These equipments will also offer job opportunities among locals (for those employed in the waste management), act as source of income to the locals (who may purchase theirs privately following the diffusion of the technology) who may have private carts for private SWM tenders and hire; and will bring in skills for managing animals and the carts themselves.

It will also develop the carts repairing and maintenance in the informal sector, thus offering a window of opportunity for self-employment. Therefore, the design should take into consideration the present and potential interests the idea of carts may invoke in the community. Training schedules for both the SWM crew and the community should be incorporated in the design. Animal feeding programmes (i.e. those pulling the carts), perhaps this may be have to be integrated with composting so that compost is used to produce livestock feed). This integrated design approach will be sustainable.

Secondary Collection: This is between the convenience transfer station and sorting point, and between the sorting site and the disposal point. Options of either 10-ton truck or and a tractor-trailer or both truck and tractor-trailer combination will be considered.

For secondary collection, some options are hereby given.

Option 1: 10-ton truck: The 10 ton truck costs over 17 million, with a very high foreign exchange component of 70% (12 million). On the other hand, a 7-ton and 10 ton tipper truck with capacities of 5 and 7 m^3 respectively cost over 14 and 19 million respectively. They have equally high force component of 65% (9 million) and 80% (16 million) respectively. Therefore whichever truck is purchased, they would be too expensive, as well as demand too much on the country's limited forex. The purchase of new trucks of whatever kind is therefore not recommended. However, should it be possible for a truck to be donated to the company, it would be a major cost saver. But its operational costs might still be higher than the cost of purchasing and operating tractor -trailer. Therefore a tractor and trailer should be purchased, even as there may exist a possibility of the company getting a donation of a truck.

Option 2: tractor-four wheeled trailer: This is a very versatile, strong and all weather option. It is durable, has a hydraulic system that it can use to get hold of and unload a trailer, and can work in rough terrains, and can easily carry the demountables from the village communal storage points every 2 weeks to the disposal point. It can also be used to collect wastes twice per week from the transfer station to the disposal point. The trailer, however, has a lower capacity of only 4m^3 compared with a 7-ton and 10 ton tipper truck with capacities of 5 and 7 m^3 respectively. However, with disposal site bound wastes comprising only 40% of total wastes, the trailer is good enough. The combined capital cost is 5 million, compared with the cheapest truck on 7 ton chassis which cost about twice.

Costing: The tractor and trailer have only 5.8 million and 0.095 million forex component respectively, giving a combined total forex component of 5.9 million for trailer and tractor. This compares more favourably with 16 million for a 10-ton truck.

Table 2: Advantages and disadvantages of using a compactor and open trucks

	1.	COMPARISON		
	Compactor Trucks		Open Trucks	
	Advantages	Disadvantages	Advantages	Disadvantages
1	Appropriate for low density wastes (It can compact)	Inappropriate for high density wastes	Appropriate for high density wastes as no compaction needed	Inappropriate for low density wastes
2	Saves labour	Capital intensive	Saves capital	Labour intensive
3	Less health hazard to crew due to less contact with waste			More health hazards to SWM crew due to more contact with waste
4	Likely to achieve payload capacity			Less likely to reach payload capacity
5	No airborne litter as body is covered			Scatters litter during transportation as body is open; can be covered using other means eg plastics.
	Good for compacting the organic matter			No compaction capacity
6	Recommended for areas with high labour costs e.g. high income countries		Good for areas with low labour charges e.g. low income countries	
7		More work-specific; less versatile	More versatile; can do a variety of jobs	
8		Very High capital cost		Lower capital cost
9		Higher		Lower

		operational cost		Operational cost
10		May require extra skills (perhaps imported) to repair and maintain	Simpler and easier to maintain and repair locally	
11		More noise pollution	Less noise pollution	
12		Imported spare parts	spare parts locally available	
13		Shorter operational life as Compaction mechanism easily spoilt by moist waste	Longer operational life as there are no delicate mechanisms; for corrosive waste, a lining can be placed to protect the body	
14		Can spoil some roads due to heavy weight	Lighter, therefore has less destructive effect on roads	
2.	GENERAL ISSUES			
1.		Generally less appropriate for the small town	More appropriate for the small town	
2.		More difficult to obtain	Easy to obtain	
3.		Load capacity greatly reduced by heavy body and compacting mechanism	Can achieve load capacity with heavy and dense construction & demolition wastes	
4.		May be unstable due to weight distribution problems	Stable, with no weight distribution problems.	

CHAPTER 7: CONSTRAINTS TO IMPLEMENTATION OF AN INTEGRATED SWM

On the whole, the general Waste management system experiences a lot of constraints. Whereas it would be reasonable to expect a comprehensive integrated waste management structure, these constraints have to be solved first. These are listed below- stage by stage:

Real and potential Constraints at primary storage level:

• Lack of awareness on the need to separate keep wastes in containers

• No previous experience of primary storage of waste, as most of it has always been thrown straight into the kitchen garden (which at times would be as close as 2 meters away from the door)

• Social and religious constraints (it may be against one's culture or religion to store waste in a container)

• Lack of awareness on the need to separate wastes at source (SAS)(if this is a proposal for the integrated system, as SAS is the most effective foundation for material recovery, and is likely to be a strong proposal for the integrated system)

• Location of primary storage equipment (outside or inside house? And method of collection (Kerbside? Block? Communal? Or door- to door; door to door collection will be tricky);

• Animals tipping the outside primary containers (e.g. horses, donkeys, dogs);

• Human Scavenging from out-of-house primary storage containers especially in estates and areas without lockable gates);

• Lack of resources for buying primary storage containers;

- Theft of primary storage containers;
- Size of container to suit different ages (children and adults) and gender (male / female) at primary collection stage;

Real and potential Constraints at primary collection level

Here the word collection is used to mean all stages involving traveling to and from the waste area or household/ yard; the collection process (waste handling loading and unloading); and delivery process: There are constraints at all of these three points.

- Container Size: to cater for all members who may be expected to carry wastes to communal storage, to Kerbside, or to waiting vehicles outside;
- Timing' (especially where block collection is planned); (there may be nobody in the house at the collection time, leading to a skip and waste accumulation);
- Security / Timing (where communal collection is done; weak residents may be waylaid and robbed/ raped especially at night or in dark streets or corners);
- Nature of container: This may be durable or delicate and as such can wear easily or not depending on who is handling it; household members may be more careful than outsiders when handling a delicate container during collection;
- Crime especially theft, robbery without violence and burglary may result from outsiders (waste collection crew, especially in door to door method) who have an access to houses when owners are at work. In some cases, some crew members or their collaborators have confused the house helps by giving them an errand, or lieing to them, and posing as

messengers from the bosses, and took away with many household items especially electrical items radios, televisions, video tapes.

Real And Potential Constraints at Secondary storage level

This section identifies and discusses some constraints met at secondary storage of solid wastes. This involves storage at the temporary transfer stations.

• Size of facility: need to be large enough to cater for the intended residents;

• Security: Some primary storage points are used by criminals for hiding;

• Sanitation problems: Some facilities used for defecating and peeing by passersby and some residents;

• Smell: Putrescible wastes decompose fast; this is bound to determine the location of facility;

• Limited resources: The most appropriate facility may be expensive, leading to a compromise facility which may have may deficiencies;

• Siting: Air-borne litter, dust, insect, rodent and smell hazards are most likely to cause objection from most residents near their houses- leading to siting difficulties;

Before implementing a comprehensive integrated SWM system, the following proposals are made to help sort out the constraints listed above. The proposals are given in figure 4-3 below.

Table 3: Ways of overcoming the constraints associated with SWM in Maun

STAGE OF SWM	CONSTRAINT	How to overcome
Primary Storage	Putting wastes in containers	Public education and awareness; Collaborative, participatory methods vital to give credibility to the system
	Diversion of use of container (a direct consequence of, and a function of poverty / poverty-related)	Offer public education; Provide waste-related income generating jobs
	Poverty / inability to pay for some services e.g. container or door-to door collection service	Design waste-related income generating projects, and have the residents have some control on who works there- with priority given to the most needy
	Social & religious constraints	Social- some extra education may be useful to enable such groups bend their rules a bit Provide services to willing buyers i.e. some tariff system varied with acceptability of level of service may be necessary; RELIGIOUS-May be as difficult as OSAMA-BUSH-SADDAM sharing a plate; may have payable services coupled with WTP[4] surveys, collaborative / participatory decision making etc Can privatize services.
	Waste separation at source	Public education; Pay households for separation service; Provide compost to complying households (those who desire; and an alternative compensation for the rest)
	Primary storage in or outside house? (In door to door collection)	Consensus building; need to ask some houses for a compromise especially; Match collection time with after-work hours
	Animals tipping wastes in outdoor primary storage	Provide containers with cover; Conduct public education on the need to cover wastes in yards; Propose a loitering domestic animal policy and assign a tax. Public education on domestic animal management and care. Shoot loitering animals on sight,

			collect and dump
	Human scavenging		Provide safer scavenging points e.g. at transfer station. Provide alternative sources of income.
	Lack of resources for buying containers		Invoke service charge and provide containers
	Theft of containers		Number the containers as per house number; Public education; Form surveillance teams in estates; Privatize service;
	Vandalism of facilities		Public education; Form surveillance teams in estates; Privatize services
	Decoding appropriate Container size		Conduct a need survey (with diagrams showing design details in the survey); Conduct participatory public workshops.
Primary Collection	Primary storage Container size[1]		Conduct a need survey (with diagrams showing design details in the survey); Conduct participatory public workshops. Let each household be responsible (this, though can cause discrepancies and confusion, though it is the easiest to do)
	Vehicle breakdown (in door-to-door or Kerbside collection)		Have alternative means especially non-motorized ones in place in the waiting list; Establish proper & comprehensive vehicle maintenance policy; Train the drivers and SWM crew on vehicle maintenance (basic mechanics of motor vehicles) so that they can sort out small failures. Offer proper supervision to ensure procedures and routes are followed properly (vehicles may break down due to overuse from unofficial errands)
	High operational costs from wear & tear of idling / waiting vehicles, and heavy fuel use during slow speed from house to house		Use non-motorized vehicles for short collection distances

---

[1] WTP means willingness to pay

	in door-to-door collection	
	Timing (especially time for door to door collection)	Consider after-work collection schedules Consensus building in public estate meetings; Can consider providing different customer-paid collection services to different houses depending on need
	Material of container (durability, weight) (Plastic? Metallic? How heavy? Etc)	Conduct house survey and demand analysis; Offer Door to door collection; Let each household be responsible.
	Handling delicate primary storage containers	Use Communal collection so that each household is responsible for the state of its primary container; Can consider using block collection; Provide durable containers at a fee;
	Street dumping	Reasonable spacing of storage bins; Form estate surveillance team (in collaboration with the residents).
	Crime by collection crew or their collaborators (e.g. daylight robbery of client residences)	Consider weekend and after-work collection schedules; Conduct a need survey to determine the extent of concern; Avoid door-to door collection;
	Insecurity / Waylaying and other related crimes especially in dark corners of the estates when wastes being taken to communal storage facilities to await communal collection.	Put sufficient lights in streets; Establish surveillance and vigilance teams within estates; Avoid communal collection system.
	Distance of storage facility from household	Let Residents decide on neutral facility site; Consensus building; Frequent and regular collection schedule; Provide enclosed storage with covers to avoid rain soaking, wind blowing, flies and smells;

		Estate Surveillance groups to monitor abuse of facilities e.g. by urinating or defecation; Clear town of domestic animals; Kill unidentified domestic animals on sight; Domestic / pet animal identity policy e.g. tagging; Fine owners of loitering animals
	Noise from motorized vehicles	Consider use of non-motorized vehicles; Fit collection time to reduce noise hazards e.g. during work hours; Minimize number of trips by using communal collection.
	Poverty	Design waste-related income generating projects.
	Health & safety dangers to the SWM crew (especially with manual handling of waste applicable to the door to door and Kerbside collection systems (waste handling risks)	Provide Health & safety training to SWM crew; Provide appropriate personal protective equipment; Use potable storage containers that can be directly emptied. Use low chassis vehicles for collection to ease placement of waste.
	Narrow and / or impassable roads	Use smaller more versatile and easy-to-manouvre vehicles such as the non-motorized ones
Secondary / communal Storage	Facility size determination	Conduct a waste generation rate survey, and project it to the next 10-30 years so that the current design caters for future expansion;
	Spacing	Conduct Survey (using contingency valuation techniques); Collaborative / partnership decision making in estate for a;
	Facility design determination	Conduct nature of waste and prevailing environmental and socio-economic status of the target location.
	Financial constraints	Make a collaborative programme with industries to whom separated wastes can be sold at a fee, also covering separation service
	Security	Public education; Form surveillance teams in estates; Privatize services
	Vandalism and other abuse of facility	Public education; Form surveillance teams in estates; Privatize services
	Environmental and health hazards (Smell, insect / vector,	Regular supervised collection schedules; Educating and awareness raising among the collection crew;

	rodent, air-borne litter etc)	Provide block collection; Provide covered storage facilities with screened walls to hide contents; Provide PPE [5] to collection crew.
	Scavenging & waste scattering	Train & engage the scavengers in the SWM work as employees; Form estate / community surveillance or vigilance teams in collaboration with the residents

Proposal For Household Hazardous Waste (HHW) Collection Programmes

To minimise improper disposal of HHW, product exchange programs, special collection days and permanent collection sites have been established by a number of communities.

Product exchange programs

Because paint products form a major portion of HHW, paint exchange programmes are being used in a number of communities to reduce the cost of HHW disposal. The reuse of latex-based paints has proven most successful, with up to 50% recovery. Unrecoverable paint must be either combusted in a hazardous waste combustor or disposed of in a hazardous waste landfill.

Special collection days

One of the most common approaches to HHW management is to hold one or more community waste collection days. On these days, the community members are asked to bring their HHW, at a little or no charge, to a specified location for recycling, treatment or disposal by professional waste handlers. In larger communities, several locations are used on successive days. Adequate promotion and education are essential for success of such programs. Records of 5-10% of such HHW are collected through such programs in Europe and it would be useful if even half of that were collected.

Permanent collection sites

To increase the convenience of HHW collection programs and therefore increase participation, more and more communities are establishing permanent collection sites (e.g. fire stations, landfills city and corporation yards etc) programs involving permanent collection facilities allow citizens to drop off wastes at their own convenience. For this reason, permanent collection sites have proven to be more effective for collecting HHW than the one-day collection programs.

CHAPTER 8: OTHER WASTE REDUCTION STRATEGIES

Source Reduction: Waste reduction may occur through the design, manufacture and packaging of products with minimum toxic content, minimum volume of material and /or a longer useful life. Waste reduction may also occur at the household, commercial, or industrial facility through selective buying patterns and reuse of products and materials. Because source reduction is not a major element in waste reduction presently, it is difficult to estimate the impact that source reduction programs have had on the total quantity of waste generated. Nevertheless, source reduction is a single most important tool for future waste reduction. Other proposals are:

- Decrease unnecessary or excessive packaging;
- Develop and use products with greater durability and repairability. This, however, goes hand in hand with capacity building and training of personnel who would do such repairs- not very friendly with most industries, but can be enforced by other mechanisms;
- Substitute reusable products for disposable, single use products such as reusable plates and cutlery, refillable beverage containers, cloth diapers and towels)
- Use fewer resources e.g. two sided photocopying
- Increase the recycled material content of products
- Develop rate structures that encourage generators to produce less waste

Recycling: The extent of recycling within a community definitely affects the quantities of wastes collected for further processing or disposal.

Public attitudes and legislation on waste generation: Along with source reduction and recycling programs, public attitudes and legislation also significantly affect the quantities of waste generated.

Public attitudes: Ultimately, significant reduction in the quantities of solid wastes generated occur when and if people are willing to change – of their own volition- their habits and lifestyles to conserve natural resources and to reduce the economic burdens associated with the management of solid wastes. A program of continuing education is essential in bringing about a change in public attitudes.

Legislation: Instituting laws for packaging can greatly help in waste reduction. Encouraging the purchase and use of recycled materials by allowing a price differential (typically 5-10%) for recycled materials is a good viable method of encouraging waste reuse. However, at the moment, the Botswana environmental and waste management legislation is seriously defective, deficient and insufficient.

Alternatives to EIA: An alternative to the project-based environmental Impact assessment (EIA) may be to institute measures such as strategic environmental assessment that involve an analysis of policies, plans and programs (PPP) at different stages of governance with a view to making them environmentally sensitive. SEA follows the same steps as the project-based EIA, except it focuses on the PPP. It views any of the P's as a project on a wider scale, and takes it through all key stages of the formal EIA. Since all these P's operate at a higher hierarchy than the project, they incorporate cumulative impacts as well. Analytical Strategic environmental assessment is a variant of SEA, but focusing on the decision-making phases. It considers decision making as a vital hierarchical stage that should be comprehensive, transparent, timely, participatory and credible. Thus it emphasises the need for wider participation of the publics prior to key decision making can be done by any party.

The adverse impacts of waste management are best addressed by establishing integrated programs where all types of waste and all facets of the waste management process are considered together. Despite their importance, limited resources may prevent these programs from being implemented, and only a piecemeal solution may be possible. However, the long-term goal should be to develop an integrated waste management system and build the technical, financial, and administrative capacity to manage and sustain it. Whether pursuing a holistic approach or a piecemeal one, managers should ensure that the program is appropriately tailored to local conditions and that practical environmental, social, economic, and political needs and realities are balanced. Answering the following key questions will help achieve this goal:
• Are adequate financial and human resources available to implement the policy, program, or technology?
• Is this the most cost-effective option available?
• What are the environmental benefits and costs? Can the costs be mitigated?

• Is the policy, program, or technology socially acceptable?

• Will specific sectors of society be adversely affected? If so, what can be done to mitigate these impacts?

For a detailed discussion of key objectives and issues to be addressed in municipal solid waste management strategies, see the UNDP Conceptual Framework for Municipal Solid Waste Management in Low-Income Countries listed under references in this document.

CHAPTER 9: SOLID WASTE MANAGEMENT FINANCING

Sources of Funding: Possible sources of funding for construction and operations are:

• Communal or municipal funds.

• Taxes. Problem: Incorporation within local tax systems Inclusion in local taxes will not work if tax collection is deficient, or if the transfer to management committees is not secured. This form of general taxation method also dissociates waste management costs and revenues.

• User charges (flat or graded rate). Block rate pricing could be used in solid waste— too: a low rate for a basic amount of garbage (the poor usually produce less waste) and higher rates for subsequent blocks.

• Mixed systems and water or electricity metering provides opportunities for cross-subsidies. Water metering can be compared to measuring the amount of solid waste produced (in volume or weight). Because electricity consumption is closely correlated with waste generation, fees for waste collection can be tied to electricity use and integrated into the electrical bill. The utility company may charge an administrative fee for handling such billing.

• Vending arrangements, such as: Shared private connections and sanitary blocks serving clusters of households. In this system, users pay in cash for each use. This system combines well with garbage collection depots.

Metered group connections paid for by a user group with its own group committee. This system is comparable to a community or group paying a private operator to collect solid waste in its area. In this case, the group is sold service from the municipal government at a bulk rate and determines its own systems for distribution and fee collection. The municipality can offer additional benefits—, for example, like exemption from certain local taxes, or a subsidy to buy equipment.

Concession system.
A system where local private operators of solid waste collection systems (micro-enterprises) obtain a license or concession from the local government. This may or may not involve community management.
• Local revolving funds or credit circles. However, voluntary funds, however, often do not generate enough money for effective solid waste management. Other communal funds that require a communal production base may not be effective in cities.
• Lotteries and auctions.
• Raffles, bazaars, or entertainment (such as movie showings).
• Donations from prominent individuals.
• Launching community-based organizations.

Fee Collection
Willingness to pay, combined with ability to manage, are good measures to assess the feasibility of a community-based project. A service is considered affordable when a community perceives it as valuable. While this strategy will lead to the desired level of service, is not necessarily the simplest or cheapest approach from an operator perspective.

Ways to generate more revenue from fee collection include:
• Change way of payment.
• Change tariff system to reflect:

Level of service. Different rates could be used for collection from communal collection points, curbside or house-to-house collection..

Type of users (domestic, institutional, commercial, industrial and gender). If men and women have their own sources of income and take part in financing arrangements as individuals, programs should avoid asking that the same contribution from women as is asked from men and women..

Income level.

Property value or characteristics.

Amount of waste to dispose (measured by size or weight of bin).

• Educate people on benefits and financial obligations. Use community meetings to review billing rate, fee collection plan, and encourage regular payment.

• Give fee collectors more personal benefit.

• Establish/enforce sanctions for non-payment.

• Fee collection by operators or respected community members rather than by government officials. Small user groups or operators can collect fees via house-to-house collection, via community meetings, via deposits on bank accounts, at government offices, or through payment in cash directly at waste disposal location. For women, payment at central places may be culturally less appropriate than home collection of fees. Payment on a savings account is also an effective strategy because women can make small deposits and poor people can join projects that require larger deposits or tariffs.

• Set fees with the assistance of community organizations. (See section on community based management of solid waste).

Accountability and Reporting:

Accountability and reporting are also aspects of financing a solid waste management project. Means of improving accountability are reporting include:

• Provide bookkeeping training, account books, water fee collection cards, etc., and employ teachers or women as treasurers.

• Avoid misuse of funds by requiring two or three committee member signatures of committee members, or one signature from someone with of the assisting NGO, to withdraw money from the bank.

• Sign a contract between the management committee and the community detailing rights and responsibilities, including reporting, for both parties. (See section on community- based management of solid waste.)

• Communicate financial reports through Bulletins distributed to households.

Oral reports given by the treasurer at community meetings followed by questions and answers.

Written reports on large sheets of paper and posted on walls in public places, particularly where people come to pay their bills.

Waste committee meetings dealing with financial matters and open to the community.

• Provide training in accountability to

Treasurers, on how to make simple summaries of costs and expenditures, and how to present these to committee and general user assemblies.

Committees, on how to account to the users for their performance.

Users, on their rights and how they can arrange for accountability (e.g., through statutory annual meetings and an independent audit committee for checking the books.)

Privatization:

Privatization is the gradual process of disassociating state-owned enterprises or state-provided services from government control and subsidies, and replacing them with market-driven entities. In the context of municipal services, privatization generally implies reducing local government activity within a given sector by:

- involving participation from the private sector; or
- reducing government ownership, through divestiture of enterprises to unregulated private ownership, and commercialization of local government agencies.

Private sector participation leaves municipal resources available for urban infrastructure and equipment. Privatization of urban services also can reduce the cost of public services to consumers; relieve the financial and administrative burden on the government; increase productivity and efficiency by promoting competition; stimulate the adoption of innovation and new technology; improve the maintenance of equipment; and create greater responsiveness to cost control measures.

Criteria for Privatization:

In deciding whether to privatize a specific aspect or portion of its service, a government needs to weigh the risks—political manipulation, changing environmental regulations, government tariff regulation, currency devaluation, inflation, and unclear taxation systems—against the economic benefits of private sector efficiency. The following criteria may be helpful in considering private sector involvement in solid waste management services (adapted from Cointreau-Levine, 1994):

Ease of defining outputs.
Ensure that defined, measurable outputs of the proposed service are incorporated in written performance specifications to clearly establish public and private sector deliverables. The government must have the resources and capabilities to monitor service levels and enforce penalties for noncompliant behaviors.

Efficiency.
Consider reasons for public and private sector inefficiencies, including cost accountability, labor tenure, government wage scales, restrictive labor practices, personnel benefits, inflexible work arrangements, bureaucratic procurement procedures, political limitations, and hiring and firing procedures. Assess options for reducing or removing these barriers. Give preference to plans offering economies of scale.

Capability.
Ensure that adequate government capacity exists for planning, design, construction, operation, maintenance and oversight. Evaluate both the public and private sectors for technical and financial resources, including expertise, skills and access to capital. Private companies must possess required facilities and equipment, or have a business plan that covers them. Governments must have both the capability to monitor performance and the political will to enforce contractual or license agreements.

Competition.

Ideally, a privatization plan will allow for competition between a number of private firms or between the government and a few private firms. Consider possible barriers to market entry and exit, as well as economies of scale that might limit competition. Determine if financial incentives or technical assistance would result in better performance from private firms. Ensure the government's ability and commitment to conducting a competitive procurement process.

Duplication.
Ensure that the government has the political will to cut personnel and assets when services are privatized. Balance the cost savings from reduced staff with new monitoring and enforcement costs.

Risk.
In some developing countries, commercial lenders and private companies do not want to risk their money on long-term or large-scale investments that rely on government payments. Regulatory framework must exist to protect the private sector against risks such as environmental damage, currency adjustments, inflation and political changes. Local governments must be able to generate enough revenue to meet contractual agreements with the private sector and protect against economic instabilities. Plans should include provisions for loss due to corruption (kickbacks, bribes and favors).

Accountability.
Ensure that private sector participation will not disproportionally benefit wealthy classes. Market openings should be made available to

small- and medium-size enterprises, helping to redistribute income. Government must guarantee a fair minimum wage and safe working conditions. Government should also make provisions for displaced workers, including job training and employment networking.

Costs.
The costs for public waste collection services must be well understood. Cost factors should be analyzed separately for the different components of solid waste service—collection, cleansing, disposal and transfer. Government must have detailed accounting information to determine whether private sector participation would be more cost-effective. A strategic planning and feasibility study should be conducted to know whether the technology offered by the private sector would result in lower costs. These criteria help to determine the extent to which a society is open or closed to competitive market forces, whether the procurement process is straightforward or opaque, how interrelated and transparent taxation and subsidies are, and the extent to which corruption skews the system. Moving public services to the private sector will be efficient only where competition, performance monitoring and accountability exist.

Privatization: Beneficial But No Panacea –CASE STUDY OF Dar:
Solid-waste management (SWM) in Dar es Salaam is the responsibility of the Dar es Salaam City Council (DCC). An estimated total of 1,929 tons of waste is generated daily from households, businesses, institutions and market centers. Before the decision to privatize solid-

waste collection and disposal, the City Council was only able to manage 2–4 percent of the waste generated daily.

The main reasons for this inability to manage waste collection were:
• Lack of equipment (trucks and machinery.)
• Lack of funds to replace equipment, purchase spare parts, service existing equipment and fuel them. Historically, DCC has allocated less than 8 percent of its total budget for SWM. Out of the 30 trucks and machinery donated by the Japanese government in 1987, only three were operational in 1992. In addition, the operational vehicles functioned at less than 20 percent of capacity.
• Lack of an official disposal site. The only "dump site" in the city was closed following an August 1991 court ruling in favor of residents of the Tabata area who complained of air pollution caused by burning waste dumped at the site.
• Lack of involvement of other stakeholders.
The DCC chose to try privatization to improve waste collection services. Privatization was accomplished in two phases, Phase I from 1992 to 1996, and Phase II from 1996 onwards. For Phase I, a single contractor was assigned to collect waste from 10 city wards and empowered to charge customers directly. For Phase II, four additional firms were given contracts through a process of open tendering, making a total of five contractors servicing 13 wards. The major achievements realized during the first phase of privatization included:
• Establishment of a solid-waste management partnership advised by a multi-disciplinary stakeholder working group .
• More efficient service and revenue collection. Households responded positively to the need to

pay for refuse collection. Initially, collection of solid waste improved to 70 percent of waste generated. However, this rate started to decline six months after the engagement of the private contractor, for reasons outlined below.

• 318 jobs were created for workers employed by the contractor. Also, human resources and stakeholders were used more efficiently; whereas 800 DCC workers collected only 30–60 tons per day, 318 workers under the private contractor collected 100 tons per day.

The problems identified in the first phase of privatization included:

Non-fulfillment of obligations from all parties. Under the contract, the contractor was supposed to pay the DCC the monthly costs of renting trucks, a leased depot, and the refuse disposal charges at the dump. DCC was obliged to pay revenue collection charges for the services provided by the contractor at DCC-owned premises like schools, hospitals, offices, etc. Unfortunately, neither party paid the other, and the DCC withdrew its facilities in September 1995. Also, the DCC was responsible for the public awareness campaigns among residents of the privatized area, and for prosecuting customers who defaulted on refuse collection charges (RCCs). When the defaulters were not prosecuted, the contractor's ability to collect revenue was further limited.

Lack of competition.
Using only a single contractor did not result in optimal pricing for the consumer or overall system efficiency.

Poor monitoring.

Staffs of both the DCC and the contractor were unfamiliar with privatization of solid-waste collection and disposal, leading to poor monitoring and oversight.

Lack of well-functioning management information system (MIS) to track payment information.

Problems within the contract agreement.
Some of the items within the contract were not well elaborated, such as the period when RCCs would be reviewed, how to deal with complaints by the refuse producers, how to monitor the daily operation of the contractors, and methods of arbitration. During Phase II, the daily solid-waste collection increased in the newly contracted wards. Solid-waste heaps were reduced, especially in open spaces and market places.

However, the constraints were similar to phase I, including inadequate payment of RCCs to the contractors. Preparations were insufficient to involve and raise awareness of people on the new strategies to clean the city and the responsibilities of individuals and stakeholders. Inadequate revenue collection prevented contractors from meeting financial targets. Contractors' equipment and facilities were inadequate, and they failed to meet promises to purchase replacements.
DCC was unable to provide an enabling environment for the contractors (e.g., information on residents liable to pay RCCs, an effective public awareness campaign). The contractors required close supervision, monitoring, support for planning, technical

advice and financial assistance. All households were not treated equally in all wards.

Limitations of Privatization:
To be successful, privatization of solid-waste management must contend with a variety of problems, including insufficient public awareness and little ability to generate the necessary public participation in planning, administering, or monitoring; managerial deficiencies and weaknesses in local authorities that make it hard to carry out policy reform measures; and lack of experienced and competent personnel to administer and manage the privatization process (see privatization story on the previous page). Municipal councils opting to privatize or commercialize their services often find that they need to upgrade all staff in accounting, auditing, information management, policy development and implementation to make these options work.

Although private solid-waste entrepreneurs work all over a city, most activity is concentrated in residential neighborhoods and biased towards middle- and higher-income households who can be relied upon to pay for services. Little or no private sector solid-waste collection activity occurs in low-income areas, due to inability to pay rather than lack of access to these areas. Large firms usually serve wealthy areas, while small firms generally serve a single, middle or lower-middle income neighborhood. Informal private sector waste entrepreneurs or "scavengers" operate in all areas. Although popular belief states that the private sector will field better-maintained refuse collection vehicles, this is not usually the case. Unless contracts provide incentives for the private firms to invest in appropriate equipment, firms lease second-hand dump trucks that frequently break down.

CHAPTER 10: COMMUNITY-BASED MANAGEMENT OF SOLID WASTES (CBM)

Community participation in solid waste management covers a variety of types encompasses several forms of local involvement, including:—

i. Awareness and teaching proper sanitary behavior;
ii. Contributing cash, goods or labor; and and/or
iii. Participating in consultation, administration, and/or management functions.

At the most basic level, participation might be providing separated waste to the waste can be handing over separated waste at a particular time to the waste collector or granting space to park waste management vehicles. With more greater public participation, the community can cooperate with public or private entities to set payment rates for service charges. Community management, the highest level of community participation, gives the community authority and control over operation, management and/or maintenance services benefiting its members. Community management may come about through partnership with governmental agencies and NGOs.

Community- based waste management CBM projects require institutional support and recognition in order to be successful. An integrated system - —including waste separation at the source, resource recovery, and composting of organic waste requires representation of waste pickers, and integration of the community to work with all and stakeholders, including representatives of waste pickers. Local leaders are often active in management of the service or maintain close contact with the municipality or community management body. Women and teens can play crucial roles, such as initiators, managers, operators, political activists, educators, and watchdogs for the community. Community-based management (CBM) can also address the following social and management problems:

Low participation of households.
Households may not participate in waste management programs because they may view solid waste management ......a low priority. They may be they are unwilling to participate in collection systems or in keeping public spaces clean, or they are unwilling to pay for service. Community Provisions for education, is often key to overcoming the best counter to these barriers, may be inadequate in but traditional approaches to waste management often do not provide enough for education. Community-based solutions can use preliminary research and input from the community to generate a list of desired services, appropriate incentives for households and servants, and systems for cleaning streets and other public places.

Management problems.
Problems with traditional waste management schemes include ineffective, inefficient, or unrepresentative management, as well as lack of community accountability to the community. CBM can introduce performance control techniques, share management with an NGO, adjust or by-pass an existing management committee, and provide incentives for managers, such as training and exchange visits.

Operational problems.

With poor motivation operators are poorly motivated, due to low salaries, low status and bad working conditions, operator motivation can be low, and public service may become can often be unreliable. Finding adequate space for waste facilities and equipment is another potential operational issue. Sound CBM can addresses motivational problems by involving operators in decision-making, using special group incentives, and, in some cases, by granting exemptions from municipal taxes. Operators can be officially introduced to households and provided with identity cards to improve operator status. Space problems can be resolved by lobbying municipalities and local leaders, as well as conducting media campaigns in the neighborhood.

Financial difficulties.

Public and private management plans often face financial difficulties caused by inadequate fee collection and inability to pay for service in low-income neighborhoods. CBM gives community input into plans for fee collection payments, incentives and sanctions for non-payment. Community input can also help waste management providers find lead to additional revenue- generating services.

Lack of municipal cooperation.

If waste collection between the municipal government and private operators is badly coordinated and the community may lose interest in trying to improve the waste situation. Extending service, mobilizing communities to lobby the municipality for assistance, involving local authorities, and structuring formal and informal opportunities for cooperation all improve municipal performance and community support for waste management plans and programs.

Capacity Building :
Insufficient capacity is a fundamental impediment to sound solid waste management programs in much of the developing world. Operating an efficient, effective, environmentally sound municipal solid waste management program requires building administrative capacity for government and private sector players and technical capacity for designing, operating, maintaining, and monitoring each part of the process. Often those people working in solid waste management—private sector companies, NGOs, and government entities—lack the technical and financial knowledge to operate efficiently. Training that builds human resource and institutional capacity at appropriate levels is essential. Peer-to-peer training for everyone from waste-picker to local government officials has proven effective in extending and sustaining these programs.

Environmental Mitigation and Monitoring Guidelines
In designing and operating integrated solid waste management programs:
• Minimize the quantity of waste that must be placed in landfills through elimination, recovery, reuse, recycling, remanufacturing, composting and similar methods.
• Manage non-hazardous wastes and special or hazardous wastes separately.
• Collect and transport all waste effectively and efficiently.

• Design sanitary landfills and ensure appropriate siting, operation, monitoring and closure.
• Establish sound fiscal and administrative management, privatizing operations with open competition, whenever feasible.

Integrating the informal sector
In Rufisque, Senegal, an innovative community initiative helped extend solid waste collection services to 3,000 households by employing horse-drawn cart operators, contracted to work two hours a day to collect refuse from households. The operators were free to work the rest of the time as general haulers. The local municipality is involved in all stages of the initiative—it is regularly represented at community meetings, assigns and approves collection routes, and maintains contractual relationships with cart operators. –(UNESCO, MOST Clearing House Best Practices Database. June, 2000)

Waste minimization (Reduce, reuse, recycle) principle.
Reducing the quantity of waste that must be transported and disposed of should be a primary goal of all municipal solid waste management programs. Waste should be recovered at the source, during transport or at the disposal site. The earlier the separation, the cleaner the material, and, in the end, the higher its quality and its value to users. Incentives which integrate and foster the involvement of the informal sector—itinerant collectors, microenterprises, cooperatives—can be essential to improved waste minimization. Other tips on reducing waste include: Encouraging recycling can help build capacity among local

micro-enterprises and reduce the waste handled by landfills and dumps.
• Organize itinerant collectors and publicize prices.
In cities throughout Africa, itinerant collectors recover high-value recyclable materials at residences and small industries. Organizing collectors can improve both their standard of living and the stability of the collection services. Publicizing prices can help stimulate the market and mitigate possible exploitation by intermediaries.
• Foster secondary markets.
The extent to which a material is recovered is dependent on the existence of local industries that can use the recovered material. Secondary markets to serve these industries do not always develop independently. Consider developing a program to identify and develop such markets where there is untapped demand.
• Offer incentives.
A deposit system on glass bottles has maintained a high recovery rate throughout the continent. South African beverage manufacturers also issue deposits for tin and aluminum cans, which have generated high levels of reuse.

Facilitate separation at disposal site.
When waste pickers are allowed access to disposal sites, significant amounts of material can be recovered. However, because they interfere with efficient operation of dumps and landfills, waste pickers are usually excluded from these sites, lowering recovery rates and causing severe economic hardship. Some sites provide a measure of structured access to waste pickers—at the Bisasar Road landfill in Durban, for instance, registered pickers from

an adjacent squatter settlement are allowed into the site after hours, earning US$77 per month from this activity. At all other times, armed guards restrict access to the site. Similarly, the South African Boipatong landfill limits access to 100 registered waste pickers.

Composting and anaerobic digestion. Organics make up 30–80 percent (~70 percent on average) of the waste stream in Africa, although this varies with the incomes of the neighborhood, region or country. If this part of the waste stream could be used for compost or methane production, many adverse impacts of open dumps and landfills would be reduced. Landfills would require less space, last longer, and produce less leachate.

Evaluate the possibility of composting.
Large centralized composting efforts, designed to separate the organic component from mixed waste, have almost always failed in Africa for reasons which include poor (or absent) feasibility studies and subsequent failure to meet cost recovery expectations. The city of Accra in Ghana has a successful creative variation on this theme: a co-composting plant that converts human waste sludge and solid waste to compost which is then sold to recover the plant's operating costs.
Small composting enterprises have fared somewhat better. Higher urban demand or subsidies may be necessary if composting is to become a part of integrated waste management. For example, a city could pay small composting operations for each ton of material that is diverted from landfills and base that payment on the disposal costs the city can avoid.

Backyard composting is a third option, but may be difficult to coordinate the level of effort needed for a city-level impact. In Uganda, community-based groups are experimenting with backyard composting, using the compost in a variety of ways, from conventional agriculture to producing fishpond algae as fish feed.

• Promote vermiculture treatment of vegetable food waste.
Small earthworm composting farms, operated by 5–6 people, have proven more successful than traditional composting facilities in developing countries, though they are not yet in widespread use in Africa. Vermiculture benefits from better quality control and the cultural perception that the final product, consisting of "worm castings," is derived from "clean" vegetable waste, whereas compost is derived from unclean "garbage." The final product is also more nutrient-rich than compost.
• Investigate anaerobic digestion.
Anaerobic digestion can generate a nutrient-rich slurry to be used on soil and a methane-rich biogas to be used for fuel.
Collection and transfer
As noted earlier, most African city dwellers lack regular waste collection or access to disposal services, except in the better-off neighborhoods or communities. Careful consideration of the city, climate, and culture is essential to achieving universal collection at recommended frequencies. The following general insights from international experience may be valuable:
• Use appropriate technology—regular trucks and alternative vehicles. Specialized compaction trucks are very expensive, difficult

to repair and often out of service. Moreover, compacting garbage provides little advantage, considering the density of the waste currently produced in most of the region. Regular trucks require less capital investment and are easier to maintain. They may also be better adapted to poor road conditions and can be used for other purposes if the municipality or company decides to transfer collection responsibility to others. For waste collection in hard-to-reach areas—very narrow streets, alleys, deteriorated roads—alternative collection vehicles should be considered, including semi-motorized carts, front-loaded tricycles, donkey carts, or handcarts.

• Integrate the informal sector.

Co-operatives and microenterprises are the primary users of smaller collection vehicles and can effectively collect waste from hard-to-reach areas at a low cost. Community members are generally more willing to pay for such flexible and inexpensive services.

• Build on the existing system.

Radical changes are often difficult to achieve, especially with limited political support, administrative and technical capacity, or financial resources. Develop new structures and processes as part of a strategy of incremental improvement.

• Introduce transfer activities.

Transfer activities often increase efficiency, for both small- and large-scale systems. In small-scale transfer, microenterprises or cooperatives bring waste to a centralized area for pickup by private or municipal trucks. In large-scale transfer, waste is transferred from a compactor or small truck to larger trailer trucks. Both types of transfer activities save fuel, reduce wear and tear on trucks, and shorten the amount of time spent traveling to and from the landfill. The farther the landfill is from the city, the greater the benefits of large-scale transfer. However, transfer activity is virtually unknown in sub-Saharan Africa.

• Shift to direct fee-for-service and local financing.

Most solid waste collection is paid out of tax revenues collected by national or local governments and redistributed to the municipality. Mismanagement of funds, lack of competition, and the resulting inefficiencies often result in non-payment or unwillingness to pay for services. Market-oriented systems in which residents' fees support collection and disposal services are less likely to suffer from these crippling flaws. Nevertheless, unwillingness to pay can still be a problem under such systems. One strategy for overcoming this problem, used in a number of developing countries outside of Africa, has been to link billing for solid waste collection to utility bills. Electricity consumption is closely correlated with waste generation, so fees for waste collection can be tied to electricity use and integrated into the electrical bill. After charging a small administrative fee, the utility passes the payments to the municipal solid waste department.

CHAPTER 11: DESIGNING AN EFFECTIVE SOLID WASTE MANAGEMENT (SWM) SYSTEM FOR A COMMUNITY.

Introduction:
The SWM realistic solutions were: Pit dumping/burning/ ground heaping/burning/burying and integrated management, including a combination of sorting / separation at source, composting, separation and reuse and sanitary landfilling. The assessment and elimination criteria are: hazard potential (pollution, landfill gas, accident, handling, health, traffic, odour, noise, and leachate), resource recovery, upgradability, aesthetics and volume control. The options are described below, with a view to guiding decision makers on the preferred solution. Incineration is not considered because it is expensive, while sanitary landfill, which the ToR considers an option of last resort, is considered a mandatory component of an integrated SWM system. Gasification and pyrolysis are not technologically feasible. Realistic options considered here include the integrated solid waste management system (ISWMS) and the no management option (i.e. status quo).

The integrated Solid Waste Management (ISWMS):
An effective integrated Solid Waste Management (ISWMS) requires these details because they will affect choices concerning:
(i) Method of storage (ii) The method and frequency of collection; (iii) The equipment used for collection; (iv) The size of the workforce; and (v) The method of disposal.
In addition, the properties of the waste also indicate the potential for resource recovery and

the environmental impact if the wastes are not properly managed.

Method of storage, collection and related issues:

Primary storage: realistic options

The realistic solutions to primary storage of waste include (i) store in one container which is replaced daily with an empty one by the crew; (ii) store separated wastes in different colour coded containers

(i) One 30-40 litre container for Household's mixed / commingled wastes;

This is most convenient among solid waste generators, as there is no agony of waste separation. This one-container alternative has the capacity to offer a lower level of, and therefore cheaper, service. The single container for mixed waste should therefore should be larger, with 20-40 litre capacity. It may be provided at a fee by the service provider, or it may be the households' (HHs) responsibility. If the latter is the case, it offers a chance for initiative, with the following possibilities: (i) Plastic bin with or without lid (ii) locally made basket; (iii) truck -tyre made bin (iv) reusable polypropylene sacks (v) one trip plastic sack etc. Bin from oil drum is unlikely to be used because it has better uses in water storage. Options ii and iii can be locally made. For the HHs preferring to avail and use their own containers, there may be a difficulty of emptying the bin, as the size may be too large for the crew members to carry. This may necessitate block or communal collection system whereby the HH is fully responsible for emptying the bin. There may also be lack of uniformity in collection, as some HHs may have an excuse of not being able to afford. Whereas it may be a good starting point, it is not sustainable. Therefore an option of the service provider availing standard containers per HH may be more tenable.

In case of town residents using a standardized single uniform container, the full containers are taken each morning by the collection crew who, in turn, replace them with empty ones. The crew put the wastes in their 70 litres colour coded containers in carts and take them to a convenience transfer station where they are taken to a handling transfer station by a trailer / tractor. HHs are therefore more likely to cooperate in this. The source of money for purchasing the containers may be critical, but an initial fund can be sought by the coordinators of the project, so that those HHs which cooperate in the program can eventually have it as a subsidy to their efforts, while those that default and decline to participate in the program for the containers. Replacement of lost or spoilt containers is at a fee. This is an option for town, especially those who may be unwilling to separate their wastes at source. It is, however, the only option available to the village HHs, who already use a lot of their wastes, and only a few, dry wastes remain which may need disposal.   Thus it needs not be collected regularly; only twice in a month may be useful.

However, the mixed wastes are more difficult to sort later than at the source, as it increases handling hazards. This is worse if some HHs do not avail their wastes for daily collection by the crew, in which case some may start decomposing, causing smell, flies and larvae hazards. This makes it more difficult to handle, sort, separate and reuse, and may be even more indirectly expensive than the option of investing in more containers. It has an aspect of hygiene which must be incorporated into the health and safety training, risk assessment and hygiene component of the sanitation program. It is a lower but more affordable level of service which can be upgraded later as demand for higher level of service grows.

(ii)   Store separated wastes in different colour coded containers of 15-20 litres:

This option involves giving each household 3-4 containers bearing different colour codes, in which they place the different wastes as they are produced. The containers can be for plastics, food remains, others (paper, wood, glass); and mixed wastes. The full ones are taken each morning by the collection crew who, in turn, replace them with empty ones. The crews put the wastes in their larger colour coded containers in carts and take them to communal storage point from where they are taken to a handling transfer station by a trailer / tractor. This option is considered more apt, as it gives the waste generators direct responsibility of separation, rendering subsequent stages easier. The wastes are separated when they are still fresh, making it more thorough and less hazardous. It reduces subsequent handling hazards, and facilitates reuse and recycling, as decomposable   wastes   are   unlikely   to contaminate the rest, and can be composted even if the collection is not immediate. The container for other wastes can be sorted further at the transfer station where a material recovery facility (MRF) exists. However, the cost of this option is higher, as the coloured containers must be bought.

Type of container options for household primary collection:
The container can be standard or different types as can be afforded by each household. The standard type is most convenient if a container exchange program is to be pursued. It, however, may not be afforded uniformly by all households due to socio-economic disparities. The colour codes make this even more complicated, as more containers must be availed. It may therefore be more practical to give the households costed options between one container (no separation at source) and many containers (separation done). Once they make the informed choice, the containers can be provided by the service provider at a cost to be borne in lump sum, by installments, or in kind by each household. The option of paying in kind involves offering services to the company- e.g. being part of the collection crew, working at the transfer station (doing separation/ sorting of mixed wastes, composting etc), making handcarts, or any kind employment at the company depending on training, competence, education etc.

Paying for the containers in lump sum:
This is (i) Convenient to the company as there is less (if any) service fee (ii) may not be afforded by most households (iii) a good indication of demand (iv) May be less commitment after payment, with a wrong notion that their relationship with the service provider is purely monetary (i.e. an element of arrogance may ensue as this lot may comprise the economic elite of Kadimo.

Paying in installments:
This is (i) less convenient to the company as it increases service fee (ii) may be afforded by most households (iii) a good indication of demand)iv) Creates some element of commitment beyond money-because of longer dealing with the company, some non- monetary relationship may result. This is better spirit in solid waste management, where there are many stakeholders.

Payment in kind:
This is (i) most convenient to the majority of residents, especially in the villages (ii) convenient to the company as it may then have committed workers who know they have a debt to pay (iii) may be afforded by most households (iv) a good indication of demand, due to the opportunity costs involved (iv Creates some element of commitment beyond money-because of longer dealing with the company, non-monetary relationship may result. This is better spirit in solid waste management, where there are many stakeholders.

Type of container: These may be metal or plastic.
Metal: These are likely to be durable but expensive, not locally available and heavy (and therefore less potable). The latter reason makes them not children-friendly, yet it s the children who are most likely to be the ones to deliver the wastes to kerbs, communal points etc. They can also be easily stolen, may not be easy to find in variety of colours, and may be too heavy for a handcart.   It is therefore less recommended option.

Plastic: They are less durable. However, they are likely to be light (therefore easy to carry by children and by handcarts), cheap, easy to find in a variety of colours, and less likely to be stolen. They are therefore more recommended.

Materials needed: (i) 15-20 litre plastic containers for poorer HHs; OR 15-20 litre metallic containers for poorer HHs; (ii) 20-40 litre plastic containers for some HHs; OR 20-40 litre metallic containers for some HHs.

Costing:
Option 1: Locally made baskets / bins:
Two 15-20 litre locally weaved basket without lid per HH @ 1000 Kshs = 2,000 Kshs/ 6 months x 2 = 4000 Kshs/ year.
Foreign exchange (Forex) component: 0%
Total cost per household: 2000 Kshs.
Advantage: Cheap, can be improved incrementally, made from local materials, saves foreign exchange, builds local capacity, and is culturally acceptable. Also, the service provider does not have to get involved, making the process less complex.. However, there may be need for the service provider to get involved to some degree, e.g. identifying competent local artists to make or supply the baskets for onward distribution to HHs upon payment. The baskets may need replacement every 6 months.

Disadvantages: May involve harvesting some shrubs, and gathering local fruits and flowers for dyeing twines for making color coded baskets. These may cause biodiversity problems.

Option 2: plastic bins Purchased from outside the area:

Two - 60 litre Plastic bins with lid: Capital cost: @4700 Kshs = 9400 Kshs/ 2 years..
Comments: Light to carry, but expensive, not likely to be uniformly purchased by the residents; Has 10% forex component, which renders it less appealing than option 1 above. May need to be replaced every 2 years, giving it an annual cost of 4700 Kshs.

Option3: plastic sacks Purchased from outside the area:
(i) 60 litre one trip Plastic sacks: Capital cost: 70 Kshs and a forex component of 10%
Comments: Light to carry, and cheap in the short run, but the most expansive alternative in the long run, with an annual expenditure of over 25,000 Kshs. It will also cause serious plastic waste menace, which is a serious environmental hazard. This option is therefore eliminated completely.
(ii) Two Reusable 60 litre polypropylene sack: Capital cost@ 180 - 360 Kshs.
These may need replacement every fortnight, giving 8,640 Kshs per year. This is too expensive.

Option 4: Metal bins:
Option 1: 200 litre made from oil drum @ 6,500 Kshs, with no forex component.
It has an advantage of zero forex component and durability. It can be used for even 10 years or more depending on the moisture content and composition of wastes placed in there, as well as whether it is exposed to rain or not. It may be locally available, but not to all HHs. Even those who have them have better uses for them than waste storage- e.g. water storage. It is also heavy. It is therefore not a worthy option.

Option 2: 80 litre galvanized sheet bin with lid @ 19,000 Kshs, with 70% forex component
Comments: Durable, and light, but: Too expensive and wastes forex. It is unlikely to be uniformly taken up even by the rich residents. It is not recommended at all.

CHAPTER 12: SOLID WASTE DISPOSAL

Hierarchy of waste disposal

The table below gives a detailed description of solid waste disposal options. It can be used to assess the stage where a city, street, court or estate is operating, and can be used to improve the system incrementally.

Non- Engineered Disposal options

This refers to all practices listed below numbered 1-5. They are popular because the engineered option requires capital expenditure, a reliable revenue stream and effective primary and secondary collection service. They have high environmental cost, and include fly, mosquito and rodent breeding, water pollution, air pollution from odour and smoke, and degradation of land. This often creates a negative public impression that all land disposal is offensive, leading officials to search for expensive alternatives such as incineration. . With some basic site operations like spreading, compacting and covering, the waste may be contained and some environmental health control may be achieved over burning, fly breeding and waste picking. However, environmental hazards from leachate and gases remain if the site is not fully and properly engineered and managed.

FULLY ENGINEERED DISPOSAL OPTION (SANITARY LANDFILLING)

Sanitary landfilling is a fully engineered disposal option. It avoids the harmful effects of uncontrolled dumping by spreading , compacting and covering the waste on land that has been carefully engineered before use. Through careful site selection, preparation and management, operations can minimize risks from leachate and gas production both in the present and the future. Site plans and design consider not only waste disposal but aftercare and ultimate land use once the site closes. (Ali et al, 1999). Sanitary landfill is suitable when suitable land is available at an affordable price. This option must be considered after an assurance that pollution could be controlled and human and technical resources are available to operate and manage the site.

LANDFILLS

Most of the waste in Africa is disposed of in environmentally unsound open or controlled dumps. Even using the best waste minimization practices at all stages, some non-recoverable waste will remain, making landfills necessary. The ultimate goal for land disposal should be:

I)Separate disposal of hazardous and non-hazardous materials; and II)construction of clean and properly sited landfills with diligent management, including leachate and methane controls, during operation and after closure.

Table 4: Waste management status- options and hierarchy Guidelines

level	Status	Description	Indicators
1	Waste discarded at source	No collection system operates. Waste is deposited by households in streets and open spaces as they generate it	No primary collection No functional institution responsible for SWM Scattered waste in streets and open areas Waste consumption by animals is common Burning of piles of waste
2	Uncontrolled local disposal	There is primary collection system and waste is taken	There is institutional responsibility for SWM Waste is from streets to nearby open places Waste quantities

		manually or in carts to a few disposal points. There is no secondary transportation using vehicles. Common in small towns.	accumulate Waste picking starts Waste consumption by animals is common
3	Uncontr olled city / town disposal	Primary and secondary collection is available. Waste is generally removed from the immediate environment and taken in vehicles to undesignated places away from residential areas	There is an institution responsible for SWM Waste is removed in two stages Transfer points are provided Often, vehicle drivers decide which disposal point to use Waste picking continues at all stages.
4	Semi-controll ed disposal	Primary and secondary collection is provided. Waste is generally removed from the immediate environment and taken in vehicles to designated places outside the residential areas. There is no management or equipment at the disposal site	Waste disposal options are in planning stage. Vehicle drivers transport the collected Waste to designated sites Waste picking continues at all stages.
5	Controll ed disposal	Primary and secondary collection provided Waste is generally taken outside the residential area	Engineered disposal options are in the planning stage Vehicle drivers transport the collected waste to designated sites Controls over waste picking at

		to designated sites in vehicles. There is some operational control and equipment / plant available at the site, though disposal is not fully engineered.	disposal site begins Solid waste authority owns the site Waste picking continues
6	Fully enginee red disposal	Waste is disposed of in a fully controlled manner with maximum protection to the environment. This is quite uncommon in low-income countries.	Details of planning and records are available. No waste picking

Source: Ali et al (1999) pp 7

When these conditions are met, the landfill becomes a sanitary landfill. It is recommended that the transition from open or controlled dumps to sanitary landfills be made incrementally. The following steps are suggested:

Open dumps. If open dumps are currently being used, initial upgrades can be made with little capital investment and minimal ongoing costs:

1. Construct perimeter drains to catch runoff and leachate.

2. Minimize leaching through soil by and repeating periodically (every two months is often sufficient compacting and grading. This causes rainwater run off into perimeter drains instead of soaking in. Manual labor or heavy equipment may be used (renting heavy equipment is often the least expensive option).

3. Protect the health of waste pickers and landfill staff by providing soap, water and hygiene training.

4. Regularly test groundwater for contaminants, including bacteria, heavy metals, and toxic organic chemicals.

5. Conduct a formal environmental assessment of the current site before making further upgrades. If it is environmentally sound and has adequate additional capacity, it can be converted directly to a controlled dump. Otherwise, an appropriate alternative site for a controlled dump or sanitary landfill must be located.

6. Engage the public in decision-making. Public involvement in upgrades, siting decisions, and subsequent planning is essential. Otherwise, strong opposition that delays or halts the project may develop.

Controlled dumps. To transform an open dump into a controlled dump:

1. Fence in the active face of the landfill and hire staff to monitor and control dumping.

2. Track how much waste is delivered.

3. Compact waste before or after dumping.

4. Schedule monitoring of methane gas production, landfill composition, and surface water and groundwater conditions.

5. Develop closure and post-closure plans.

6. Seal and cover the dump in stages as its capacity to receive waste is exhausted.

7. Maintain scheduled monitoring until sampling indicates it is no longer necessary—possibly 30 years or more.

Sanitary landfills.

Sanitary landfills are the only land disposal option that enables control and effective mitigation of Potential surface and groundwater contamination; Health and physical threats to waste pickers and sanitation workers; and methane emissions. Sanitary landfills require much greater initial investment and have higher operating costs than controlled dumps. Full community involvement throughout the life cycle of the project is essential. Proper design, operation and closure also require a much higher level of technical capacity.

Siting.
Siting is possibly the most difficult stage in landfill development.
1. Carry out an environmental impact assessment that addresses all siting criteria (see box at left).
2. Organize full community involvement. This is especially important given the greater expense and often greater size of sanitary landfills.

Design.
To mitigate environmental impacts, sanitary landfill designs should include:
1. An impermeable or low-permeability lining (compacted clay and polyethylene are most common in developing countries; geopolymers and asphalt are prevalent in the developed world).
2. Leachate collection, monitoring, and treatment.
3. Gas monitoring, extraction, and treatment.
4. Fencing to control access.
5. Provisions for closure and post-closure monitoring and maintenance.

Leachate management.
Leachate impacts can be controlled only with lined landfills.

1. Install collection systems to retrieve leachate from the bottom of the landfill.

2. Treat leachate physically, chemically, or biologically through:

a. An off-site sewage treatment plant (adequate sewage treatment facilities are readily available in only some parts of Africa), or in a dedicated on-site treatment plant.

b. Recirculation that sprays leachate from the bottom of the landfill onto its surface. This is a popular landfill management practice in Africa. It reduces leachate volume by increasing evaporation, stores remaining leachate in the body of the landfill, and may accelerate degradation and extend the life of the site. However, recirculation is a new technique whose long-term effects are not yet known.

c. Evaporation of leachate through a series of open ponds. This method requires pumping and some means for disposing of possibly toxic residues. Ponds should be designed with enough capacity to accommodate increased volume during the rainy season.

3. Monitor groundwater and surface water regularly, both down-gradient and up-gradient from the landfill. At a minimum, monitoring should include indicators of core contaminants, chemical oxygen demand, biological oxygen demand, and total nitrogen and chloride levels.

4. If it is uneconomical to recover and use landfill gas as fuel, it should be vented and flared. Currently, recovery and processing systems are both expensive and difficult to operate. These systems are economical only when the landfill generates large quantities of gas, where local or regional demand exists, or where the price for natural gas or other substitutes is high. At a minimum, buried perforated pipes that can safely vent gas should

be installed, and a flaring system should be added to reduce global methane release to the atmosphere.

5. Fence in landfills to prevent waste pickers from accessing the site. This enables landfill personnel to work efficiently and protects waste pickers from exposure to harmful substances. However, it also deprives them of their livelihood. They should thus be integrated into formal collection or disposal operations by, for instance, helping them organize a cooperative and offering them structured access at the landfill gates. Also, they should be made a part of the earlier stages of the collection process, perhaps by helping them establish a cooperative that collects recyclables from industry.

6. When the landfill is full, implement the activities specified in closure and post-closure plans that were developed during design. These should include sealing the landfill and applying a final cover (including vegetation) to it, land use restrictions on the old landfill and surrounding areas, and long-term gas, leachate, surface water and groundwater monitoring.

Siting guidelines for landfills
Do not site landfills:
• In wetlands or areas with a high water table
• In floodplains
• Near drinking water supplies
• Along geological faults or seismically active regions
• Within two kilometers of an airport
Do site landfills:
• Above clay soils or igneous rock
• With active public involvement
• In areas with sufficient capacity

CHAPTER 13: SANITARY LANDFILLING / FULLY ENGINEERED OPTION

Sanitary landfilling is a fully engineered disposal option. It avoids the harmful effects of uncontrolled dumping by spreading, compacting and covering the waste on land that has been carefully engineered before use. Through careful site selection, preparation and management, operations can minimize risks from leachate and gas production both in the present and the future. Site plans and design consider not only waste disposal but aftercare and ultimate land use once the site closes. Sanitary landfill is suitable when suitable land is available at an affordable price. This option must be considered after an assurance that pollution could be controlled and human and technical resources are available to operate and manage the site. Ali et al. (1999) state that hazards arising from landfill can vary from one site to another, but depend primarily on a range of factors including waste composition, moisture and climate. As a guideline, sanitary landfilling should meet the minimum requirements as set below:

SLF Location
• Careful siting to minimize groundwater and other potential pollution problems;
• Ideally sited away from present and proposed residential areas but not to the extent that transport costs become unaffordable;
• Adequate barriers to protect nearby residents where present;
• Control of wind blown litter (paper, plastics etc) by screening and cover.

SLF Operation

1. Minimize contact between waste and water;
2. Compaction of refuse and covering to prevent nuisance through flies and vermin;
3. Prevent the formation of pools of water where mosquitoes could breed;
4. Discourage rodents and enable early discovery of burrows through monitoring;
5. Minimize smells and prevent burning by compacting and covering refuse and controlling site operations;
6. Fill depressions so that profile is uniform;
7. Control birds through prompt covering of waste.

SLF Management and control
• Make site ownership and responsibility clearly identified;
• Earmark site officially and ensure it as actually used by persons allowed at the site;
• Actively monitor and control site operations;
• Restore the site to an acceptable condition after closure;
• Plan for future use of the site (Ali et al, 1999)

Preparation for and Designing of a landfill site;
A sanitary landfill is a fully engineered facility for safe handling and disposing of solid wastes. Design basis will be incremental upgradability of the existing system to ascertain sustainability, taking cognizance of the need to gradually develop the project while developing the support capacity in readiness to operate in a fully engineered disposal facility in the longer run. The intermediate stages would involve establishing and organizing a dumping ground with a transfer station where waste sorting with a view to reduction, reuse and recycling is

done. This would run alongside increasing rate, level and efficiency of the entire SWM system that eventually feed the proposed landfill. Eventually, as the wastes become more potent with increasing industrial development, a sanitary landfill will be established, and it will automatically fit into a well running system. When it comes, the following will be the guide on selection of type of landfill.

Selection of landfill type:
There are three main types of landfills according to the state of California (1984) (Tchobanoglous et al 1993). These are: (i) Conventional landfills for commingled MSW (ii) Landfills for milled solid wastes, (iii) Monofill for designated or specialized wastes (iv) Others e.g. (a) maximum gas production system; (b) integrated solid waste treatment units (c) wetland landfills. In places without appropriate cover material, e.g. where all soil is sand, it would require that appropriate material to be imported from elsewhere. To reduce this expense, it would be better to consider seriously the integrated solid waste treatment units. This involves the organic constituents being separated out and placed in a separate landfill where the biodegradation rates would be enhanced by increasing the moisture content of the waste, either by recycling leachates or by seeding with digested wastewater treatment plant sludge or animal manure. The degraded material would be excavated and used as cover material for new fill areas, and the excavated cell would be filled with the new waste. This can give additional landfill capacity.

Other Salient Features Of A Functional Sanitary Landfill

The development of a workable operating schedule, a filling plan for the placement of solid wastes, landfill operating records and billing information a load inspection plan for hazardous wastes and site safety and security plans are important elements of a landfill operation plan.

Sanitary Landfill operation:
The standards for sanitary landfilling are flexible yet enforceable. The following is a profile of the steps and activities followed and required for proper daily running of a SLF

(i)   Wastes will be placed in compacted layers of not more than 2 meters deep. This will help in proper packaging of the waste, and will save space and extend life of landfill.

(ii)   No loose and hollow litter will be allowed on site, as they can contain water, or hide undesirable animals. Tyres will be shredded, or recovered at the transfer station for recycling. If they have to be land filled, they must first be filled with other compact, firm, solid stuff. This will be to reduce vector, insect and rodent menace.

(iii)   Special wastes (e.g. hazardous wastes – radioactive, infective clinical, special industrial wastes etc) must be disabled (e.g. pre-treated) by the generators prior to disposal. Properly trained and appropriately remunerated officers will be at the inspection site to check on this;

(iv)   Only Sludge with less than 50% moisture content will be allowed. This is to reduce Leachate volume, and thus help reduce pollution potential of the Leachate.

(v)   Only specific types of vehicles will be allowed at the site; the rest can take their wastes to a transfer station. This will ease traffic control at the site, thereby reducing accidents, and reducing health and safety hazards to the human traffic.

(vi)   No scavengers will be allowed in site, except those with special permission e.g. those working together with the private companies contracted to the convenience transfer station at the landfill site. This will be to facilitate landfill operation activities, including reducing vehicular accidents.

(vii)   Vehicles will not be allowed to drive at more than 20 km/hour while within the SLF premises. This is to help reduce dust menace and vehicular accidents.

(viii)   Loose and low-density waste (density less than 200 kg/m^3) will first have to be bailed (or other appropriate densification method) before being allowed.

(ix)   No person will be allowed on site unless they put on the necessary personal protective equipment. These will include overalls, boots, head and face protection, nose protection (respirator). This will be to reduce exposure to health and safety hazards.

(x)   Each load will be accompanied with a short description of its general nature and contacts of source. This is to facilitate quick assessment at the SLF gate to reduce traffic congestion.

(xi)   Each load will be spread, compacted, and each cell given a daily post-compaction cover of not less than 15 cm. This is to reduce litter scattering, control smell, rodents and vectors.

(xii)   There will be a potable fence (Chain-link) placed around the working face of the landfill to control littering (especially air-borne). A patrol team will pick the litter sticking on these fences every day.

(xiii)   The   quantities,   source   and approximate characteristics of incoming wastes will be recorded at the SLF gate.

(xiv)   All vehicles coming to the SLF will have their wastes covered. This is to reduce smell and air-borne littering.

(xv)   There will be no smoking at non-smoking zones of the SLF site. This is to control fire hazard, as the SLF has lots of combustible wastes that can easily catch fire, and later, landfill gas increases the fire risks even more.

Load inspection:

The process of unloading the contents of a collection vehicle near the working face or in some designated area, spreading the waste out in a thin layer and visually inspecting the waste to determine whether any hazardous wastes could be present. The presence of a hazardous material can be detected by a hand-held radiation-measuring device or at the weight station. If hazardous wastes are detected, the collecting company is responsible for removing them (polluter pay principle). In some cases, if the company brings such material a second time, heavy fine is levied. If it brings a third time, it is banned from discharging wastes at the landfill. There should therefore be a means of monitoring the quality and quantity of waste at the landfill site.

Landfill closure and post-closure care:

These involve what is to happen to a completed landfill in the future. There should be a budgetary provision for maintaining the closed site into perpetuity, mostly 30-50 years into the future. The closure plan must include a design for the landfill cover and the landscaping of the completed   site,   and   long-term   plans   for controlling runoff, erosion control, gas and Leachate collection and treatment, and environmental monitoring.

Cover and landscape design:
These should be a plan to restore the landfill site to its original, if not better state. This is done by beautification procedures (landscaping), which may involve the use of landscape materials such as hard paving (bricks, slabs etc), soft paving (sand, crushed stones etc), waterfront (pond, fountain) or plants (trees, grass, shrubs or ground covers). A combination of these can be used to ensure the site is attractive enough and poses no esthetic pollution. Wherever possible, the scraping and stockpiling of native topsoil for later use as the final cover for the closed landfill is recommended. This is particularly advantageous when the end use is the restoration of the site to its natural condition and native plants are to be used. These need the availability of local soil, which reduces stress factor for plants growing under inherently adverse conditions of a closed landfill site.

Control of landfill gases:
Landfill gases must be controlled for as long as they are expected to be generated after the landfill is closed. This may be by use of extraction wells, collector and transmission piping, and gas flaring and/ or combustion facilities. A means of monitoring, collection and management (e.g. by flaring, or re-use) of landfill gas (LFG) should be in the landfill design right at the conception stage. The LFG management should be done throughout the design, construction, use, closure and post-

closure stages to avoid fire and air pollution hazards.

Collection and treatment of Leachate:
Landfill Leachates are liquids washed from the landfill wastes. They largely comprise organic acids, and have high Biochemical Oxygen demand (BOD) and Chemical oxygen Demand (COD). Leachates therefore have the capacity to contaminate groundwater, but are also able to transport dissolved organic substances that may be released in the unsaturated subsurface environment, by the change in the partial pressure of the constituents in the gas phase. There should be a proper Leachate collection and monitoring strategy throughout the entire life of the landfill (i.e. from conception to post-closure). Their characteristics change with age of the landfill, and these changes should be monitored to be sure they are steady. Any major deviation from normal pattern should be studied and monitored more closely, as it may indicate an interaction with other media such as ground water. This is best done by environmental monitoring systems.

Environmental monitoring systems (EMS):
The Environmental monitoring system (EMS) is necessary to ensure that the integrity of the landfill is maintained with respect to the uncontrolled release of any contaminants to the environment. In most cases, the selection of the facilities and procedures to be included in a closure plan depend on the environmental control facilities used during landfill operation before closure. Designers should chose monitoring facilities that can be used to track the movement of any landfill emissions to the water, air and soil environments. This involves

regular check on the characteristics of the LFG, leachates and vadose zone. This should cover the monitors in the vadose zone, water wells and well caps, gas probes and survey monuments. Therefore vadose zone monitoring involves both liquids and gases. Liquid monitoring in the vadose zone is done using suction lysimeters.

SLF Post-closure care:
The type of care should depend on the use of the site. Closed landfill sites can be used as recreational areas, parks, nature preserves, botanical gardens, crop production, and commercial development. Each use presents its own unique challenge. The facilities at a closed landfill must be maintained over the period of time that the landfill is producing products of decomposition. This ranges mostly from 20-30 years, but 50 is also possible the key issues to be addressed in a landfill closure plan are:
• Routine inspections
• Infrastructure maintenance (Grading and landscaping, Drainage control systems, Gas management systems, and Leachate collection and treatment)
• Environmental monitoring systems
• Reporting
• Facility changes
Emergency response plan

Routine inspections:
These are done to characterize the condition of the landfill closure facilities.

Infrastructure maintenance:
The infrastructure of landfills includes grading and landscape features, drainage control systems, gas management systems, and

Leachate control systems (Tchobanoglous et al 1993). This infrastructure must be maintained systematically through a planned schedule of preventive maintenance to protect the integrity of the landfill cover and prevent contamination of air, water and soil environment adjacent to the landfill. Funds and equipment for these activities must be put aside. Special attention should be given to the landfill cap repair, which includes a geomembrane liner. Both runoff and run on surface waters must be controlled. It may be necessary to install and operate storm water pumps after many years of landfill settlement. Maintenance of drainage control systems must be coordinated with maintenance of land surfaces and revegetation of landscape plans.

Environmental monitoring systems:
This should cover the monitors in the vadose zone, water wells and well caps, gas probes and survey monuments. Vadose zone is the space between the soil surface and the permanent ground water level. Vadose zone pore space is not filled with water, and the small amounts of water appearing there are mixed with air. Therefore vadose zone monitoring involves both liquids and gases. Liquid monitoring in the vadose zone is done using suction lysimeters.

Types Of Landfills
There are three types of landfills according to the state of California (1984) (Tchobanoglous et al 1993). These are:
(i)   Conventional landfills for commingled MSW
(ii)   Landfills for milled solid wastes, and
(iii) Monofill for designated or specialized wastes

(iv)   Others e.g. (a) maximum gas production system;   (b)   integrated   solid   waste treatment units (c) wetland landfills

CHAPTER 14: MATERIAL RECOVERY, PROCESSING AND REUSE

It is expensive to design and construct waste management infrastructure. A landfill may const billions of money. The idea therefore would be to use it for as long as is possible. For this to happen, what is deposited there must be carefully planned and selected. This involves a lot of material recovery from the waste stream prior to final disposal. This is the essence of material recovery facilities (MRFs). Whereas MRFs are formal components of material recovery, recovery takes place at all levels of the waste chain, through the informal networks such as waste pickers and scavengers.

Whereas picking occurs at all levels, waste picking and scavenging are most common especially at the dumping grounds. Mostly commercial and garage wastes, Food and other non-organic materials picked. There is no provision for Health and safety as scavenging is done with bare hands and inappropriate (or no) footwear. They also do it alongside vehicles at the dumpsite. This exposes them to hazards from vehicular accidents, vehicle exhaust fumes, noise, poisoning and infections from the foods, and contagious diseases from contact with infected and toxic wastes. There should be control over this habit, with a central site for waste picking provided. This can be at a safe site within a transfer station (which currently does not exist, but can be included in the currently planned Sanitary Landfill (SLF), and a legal requirement for all waste handlers, including those visiting any SWM sites, to be in full PPE gear, enforced, followed by credit / financial schemes to acquire these equipment.

The waste pickers can be trained, and organized into cooperative societies so that they are more organized, and can operate on a more-business-like, respectful, and formal level. The idea of getting them alternative livelihood may not be practical, as joblessness is rampant in Botswana. The informal sector (of which they are already a part) can be organized and expanded to cater for more participants in a more organized way.

Current Recycling programmes and other available recycling opportunities
Recycling is the primary use of a waste (in its intact form) or secondary use of whole or part of the waste after minor or major modifications (Ali, 2003). Recycling reduces the quantity of waste to be disposed of in a landfill, and is one of the most environmentally friendly waste management methods. The re-used items would gradually reduce in volume through wear and tear (e.g. old clothes used ad duster, mattresses, brooms etc.

Waste Composting:
Meaning of Compost:
Compost is the humus-rich, decomposed organic matter. It may be produced by aerobic or anaerobic process. Aerobics composting is more popular and common because the final product has no objectionable smell.

Methods of composting
Among the most common composting method is the windrow composting Ali, 2003). This involves arranging the organic waste in rows and either aerating by passing air or blowing used air from it. This replenishes the oxygen

supply to the decomposing matter. Regular turning of the waste is necessary to ensure that all waste get a chance of being at the centre whose temperature of 700C kills most weed seeds, pathogen eggs, parasites and other undesirable living organisms in the solid waste.

Vermicomposting

Worms can be introduced into the organic matter to decompose it faster. The worms eat up the organic matter and throw a solid waste called castings. This is rich in many minerals that plants require for normal growth. Decomposed organic matter (hereafter called compost), is rich in a number of macro and micronutrients.

Importance of compost in Agriculture

Compost is mostly used in farms as soil conditioners. It adds both macro and micro nutrients to the soil; it provides food for soil microbes- thereby making the soil healthy; it helps stabilize and build up the soil structure (mostly produces a crumb structure); it improves soil moisture and nutrient holding capacity; it improves drainage and aeration of the soil; and it cushions soils against extreme temperature changes (since it is spongy and therefore has air which is a bad conductor of heat). Therefore, it is vital that any compost produced should be able to perform the above roles effectively. This is why marketing and understanding of demand come in.

Demand for and Marketing of Compost

Compost is made from organic matter- rich solid waste. The composition of the waste should therefore be of the right standard if there is to be a growing demand for it. Secondly, before producing compost, one should know the target market and demand - and ideally, there should be market for the product. If there is no market, but raw material, equipment and skills for composting are available, then one should first do marketing to create demand for compost. This can popularize the compost so that the venture becomes viable. Since farmers already have popular alternatives to the organic fertilizers in the form of artificial / chemical / mineral / inorganic fertilizers, the demand for compost should be gauged first before it is produced- especially if farmers are the target market. The farmers should be made to understand the multi-dimensional role played by the organic manures- a means of sustainable production. They should understand the ideal composition, and the product supplied should meet the standards. Before full-scale production is initiated, there should be a pilot project to gauge the viability- otherwise another white elephant project results.

Disadvantages of Compost
(a)Bulky (b) has only small concentrations of plant nutrients (c) Slow acting, i.e., releases nutrients only very slowly. (d) Labour intensive (okay for low income countries where there is cheap and reliable labour). However, the compost releases its nutrients for a long time, besides the role of conditioning and improving the impervious clay and over-drained sandy soils.

Characteristics of good Compost
Compost should ideally have the following characteristics (Ali, 2003):
(i) C:N ratio of 15:1 up to 20:1
(ii) Moisture content not more than 35 %

(iii) Have organic matter level of at least 25%

(iv) Should not have large amounts of toxic matter such as heavy metals;

(v) Fine compost Should be able to pass through a 18mm sieve; medium compost should be able to pass through a sieve of 40 mm;

(vi) Should not have any smell, i.e., should be stable;

(vi) Should be fully decomposed.

(vii) Should not have any undesirable objects such as broken glass, pieces of metal etc.

These parameters describing ideal compost can be measured by the methods described below :(Adapted from Ali, 2003)

Table 5: Characteristics of a good compost

Item	Parameter	Ideal level	Method of Measurement
1	Moisture content (MC) (in %)	<35%	(i) Get weight of a pan (ii) Get weight of fresh compost in the pan (iii) Dry the compost in an oven at 105^0C (iv) Cool, then get weight of dry compost and pan. Difference between ii and iv gives the MC. i.e., the MC is the difference in weight between the fresh weight of compost and the oven-dried weight of the same amount of compost.
2	Organic matter (OM) content (%)	>25%	(i) Weigh cooled oven-dried dry compost (ii) Completely burn the oven-dry compost and then let it cool (iii) Weigh the residue. Organic matter content of of compost will be the difference between (i) and (iii), divided by (i), then multiplied by 100 to get %.
3	Toxic substances e.g. heavy metals)	Very low	
4	C:N ratio	Between 15:1 and 20:1	
5	Smell	None	Smelling

6	Fully decomposed	No intact organic matter	Visual
7	Diameter	< 18 mm for fine compost < 40 mm for coarse compost	Placing intact compost in 40 mm sieve, and shake. >99% should pass through. Then place the sieved compost again on an 18 mm sieve, and shake. A good proportion should pass through. However, if it is meant to be all fine compost, then all (OR > 99%)should pass through an 18 mm sieve when shaken.
8	Other	No undesirable materials e.g. pieces of glass, metals etc	Feel by the thumbs / fingers; also visual

Other possible uses of Compost

The alternative uses of the compost can be considered. These may include use as landfill cover, heat-treatment to release energy (e.g. pyrolysis or gasification), etc. In places where there is no appropriate landfill cover e.g. sandy soils; composting can be an integral part of the landfill design so that a compost plant is installed right at the landfill.

Advantages of Composting in SWM

Composting supports solid waste management because it reduces the amount of wastes that could otherwise have occupied space in the landfill; it reduces the landfill gas menace- as most organic matter which would release such gases are not land filled; it facilitates the material recovery process; and its production helps lengthen the lifespan of a landfill.

For a proper material recovery process to be instituted, a proper analysis of the waste stream is a prerequisite. The section below gives a brief guide of waste sampling.

Waste sampling:
Quartering:
This is a method of ensuring as representative a sample as possible is used to get the necessary data. This ensures reliability and replicability of the data. Quartering involves the following steps:
(i)   Place a plastic bedding flat on flat, form ground;
(ii)   Open up and empty all containers with wastes onto this plastic;
(iii)   Mix well using any appropriate method;
(iv)   Separate the mixed waste into two equal halves along a line (Made of either string or a metal or plastic plate;
(v)   For one of the halves from (iv) above, separate it further into two halves (thus giving a quarter of the original sample);
(vi)   Remove three of these portions, remain with one on the sheet, and repeat the processes (i) – (v) above until you have the correct amount / waste size as per the guideline in column 3 of the table above.

CHAPTER 15: RISK AND ENVIRONMENTAL ASSESSMENT OF SANITARY LANDFILL (SLF)

By 2001/3 period, many Botswana cities were planning to establish sanitary landfills. Taking que, the North West District council, Maun, Botswana (NWDC) had planned for a fully engineered SLF in Maun and other towns in Botswana. This was ostensibly to cope with the then existing logistical challenges of SWM. Besides the clear unpreparedness of Maun's NWDC to operate a SLF, there were also inherent risks and hazards, which needed to be addressed at different stages of the project cycle. An assessment done in Maun in 2003 indicated that garbage collection rates were below 60%, with even storage at primary and secondary levels facing managerial challenges. There were inherent dangers at public health and environmental level. This chapter discusses some pertinent issues relating to this facility. They mostly relate to environmental and public health. Whereas SLF remains the top most disposal method, it requires a well organised system at lower levels which feed it. Even after establishment, it has some inherent environmental and occupational challenges associated with it. These are the subject of this chapter. They range from explosion of LFG, LFG pollution as an active climate change agent, leachate pollution, roll over, vehicular accidents, gas leaks, odours and aesthetics, among others discussed below.

Explosion:
Both methane and hydrogen are flammable in the presence of oxygen and potentially explosive in a confined environment. Methane is flammable in air within the range of 5-15% by volume, while

Hydrogen is flammable within the range 4.1-7.5%. When fire strikes, there is almost always an explosion. However, hydrogen is seldom present at levels within the explosive range. Where the escaping gas is not confined, the explosion risk is lower but there is a risk of fire (Ali, 2003). (i)Fires: These may result from (i) hot ashes in a vehicle delivering waste, (ii) a cigarette end thrown by a worker, (iii) Sun's rays through a fragment of glass on the surface (iv) any others such as electricity, etc. Both methane and hydrogen are flammable in the presence of oxygen and potentially explosive in a confined environment. Methane is flammable in air within the range of 5-15% by volume, while Hydrogen is flammable within the range 4.1-7.5%. However, hydrogen is seldom present at levels within the explosive range.

Vehicular accidents:
May be common especially where human movement (scavengers) is not controlled.

Displacement of people:
This may cause some inconveniences to the affected families.

Roll-over:
These are accidents likely to arise from the large vehicles such as compactor, tractors, trucks and lorries manoeuvring at the site either do deposit / unload waste, spread waste or compact waste. Refuse truck drivers face the risk of rollover especially at the access ramp when entering or leaving the site.

Smoke and toxic fumes
Poisoning from fires may lead to asphyxiation and death, or other levels of poisoning arising from

toxic combustion products such as carbon monoxide and Sulphur dioxide.

Oxygen deprivation
This affects plant roots due to gas migration, leading to vegetation dieback. Where gas migration has occurred, the pathway is often indicated by surface vegetation, including trees, which show withering at leaf margins, defoliation and branch dieback.

Trace components:
Trace components of LFG comprise mostly alkenes and alkanes, and their oxidation products such as alcohols, ketones and aldehydes. Many of these are recognised toxicants when present in air at concentrations above the established toxicity threshold limit values (TLV) or Occupational exposure standards. However, in most working places, significant dilution occurs through mixing with air, reducing levels to safer limits. It is nevertheless important to recognise the hazards and potentially dangerous situations related to these trace components (Ali, 2003).

Global warming.
Green house gases such as carbon dioxide, methane, and CFCs accumulate in the atmosphere and form a blanket that absorb reflective solar radiation, thereby blocking their release from the atmosphere. This blanket, however, freely allows incoming solar energy to pass into the atmosphere, leading to elevated temperatures (also called global warming). This is currently a top global concern.

Leachate poisoning:

Leachate is the liquid that has passed through a landfill, and contains some chemically active and toxic dissolved materials acquired from the landfill waste material. The key leachates are of concern due to their high Biochemical Oxygen demand (BOD), Chemical Oxygen demand (COD) and / or heavy metal contents. The BOD and COD rich leachates can cause ecological imbalances in surface and ground water resources such as eutrophication, and lead to blooms and their inherent algal toxins. The heavy metals may cause a wide range of health risks, including genotoxicity, nervous imbalances and bioaccumulation in the ecosystem.

Ammonia:
The presence of ammoniacal nitrogen means that Leachate often has to be treated offsite before being discharged, since there is no natural biochemical path for the removal of this constituent (Ali, 2003)

Odour hazard:
The gases may produce bad smell to the surroundings, and these may affect workers. May affect both site workers and the surrounding residents.

Dust and fines:
These may cause asthma, respiratory diseases and visibility problems around the site. Affects both site workers and residents.

Aesthetic(s)
This includes concerns arising from scattered waste around the site. This may lead to

untidiness, health hazards, and related management and public perception problems.

Blowing debris (also related to dust and fines). May affect site operatives and surrounding residents.

Insect vectors,
Vectors, rodents and diseases such as bubonic plague from the rats, and insect vector-borne water-related diseases e.g. malaria. May affect both site workers and surrounding residents.

Gas leaks:
Whatever precaution is taken, some gas may leak into the surroundings and cause a risk to adjacent buildings. There may also be smell from these gases.

Unsightly depressions:
These may arise from sites where cover and capping material is mined. Pools of water may collect in these, causing health problems to both site workers and neighbouring residents.

Figure 1: Routine Inspections of landfill closure facilities

Inspection item	Frequency of inspection	Potential problems to be observed
Final cover	Once per year, and after each substantial rainfall	Erosion to expose the synthetic liner; landslides
Vegetative cover		Dead plants
Final Grades	4 times / year	Standing ponds of water
Surface drainage	Twice / year	Debris in drains; broken drain pipes
Gas monitoring	4 times/year, and after each substantial rainfall.	Odors, compressor and flare equipment inoperable; high gas readings in monitoring probes; broken gas well pipes.
Groundwater monitoring	Continuous as required by the post closure maintenance & management plan	Damaged wells; inoperable sampling.
Leachate monitoring	As required by the post closure maintenance & management plan	Inoperable Leachate pumps; blockage in Leachate collection pipes
	As required by the post closure maintenance & management plan	

CHAPTER 16: MANAGEMENT OF THE ENVIRONMENTAL AND HEALTH IMPACTS FROM LANDFILL TECHNOLOGY

This section makes proposals on how best the environmental and occupational impacts associated with SLFs, as well as associated risks and hazards discussed above can be addressed at different stages of the project cycle.

Correct Personal protective Equipment (PPE):
Staff in the landfill site should be issued with appropriate PPE sufficient for their particular job. Items should include:
a.   Safety boots;
b.   Protective overalls;
c.   Hand gloves (sometimes puncture-proof gloves may be necessary);
d.   Safety glasses;
e.   Hard hats (helmets);
f.   Dust mask
g.   OR Air-filtering head gear which covers items d, e and f together.

These should be measured to fit each user, so that they don't become the barrier to health and safety. On top of these, there should be suitable laundry arrangements for keeping these items clean and safe.

Limiting production
This may be accomplished by use of physical barriers to limit the migration or by reducing the pressure within the landfill using vents, which provide a controlled path of least resistance for the gas. This may involve laying barriers around the edge and base of the landfill. These increase vertical flow, and reduce lateral flow.

Pressure reduction within the landfill:
This may be accomplished by (i) passive control-where pathways are created and the gas extracted solely under the power of the pressure differential. This collects less gas, which cannot be used; or (ii) active control whereby gases are sucked from the landfill. This can be done by positioning wells or gas vents (network of collection pipes) throughout the landfill structure, or stone-filled gas vent trenches placed around the perimeter of the site. This reduces the pressure and the chances of LFG migrating from site. The gas can be flared or used for energy reclamation purposes. Flaring means burning to avoid its ingress into the environment and to reduce its harmful effects. This may be done using a small gas burner at the point source (i.e. where the collection system meets the atmosphere).

Fire and explosion:
By controlling pressure within the landfill, explosion can be avoided. However, fire needs to be approached from the perspective of the fire triangle. Care needs to be taken to avoid a combination of these three, especially source of fire. This is because by its very nature, a landfill generates fuels such as methane, Hydrogen and carbon monoxide, as well as being exposed to the atmosphere. The latter means there is enough oxygen. What lacks in the combination is source of fire, which can be controlled if appropriate care is taken. Staff should be trained in fire management, through fire drills etc. Some reliable source of water for immediate use in fire control should be availed.

Gas leaks:

Whatever precaution is taken, some gas may leak into the surroundings and cause a risk to adjacent buildings. This can be contained by monitoring such migration. Planning gas control on target basis can make environmental monitoring more efficient. This means a possible target is identified, and a gas control system developed to protect that structure. This can allow the safe construction of buildings close to a landfill site.

Environmental monitoring:
Even after landfill site is closed, gas and Leachate production continue for a long time. Therefore any changes in quantities and composition of both Leachate and LFG should be noted and monitoring systems to detect the release of either into the surrounding environment should be provided. Monitoring facilities are required at new landfills for (i) gases and liquids in the vadose zone, (ii) For groundwater quality both upstream and downstream of the landfill site, and (iii) for air quality at the boundary of the landfill and from any processing facilities (eg flares). Sampling points can be constructed surrounding the site to monitor any changes in the gases present in the ground water, the vadose zone and atmosphere, as well as groundwater quality changes.

Airborne litter:
Some waste may be scattered around especially on windy days, leading to untidiness, health hazards, and related management and public perception problems. Litter control can be accomplished by construction of portable litter fences around the working area. They can be easily moved as the working area expands, as the working face shifts, and as the wind changes

direction. These should be cleaned daily by manual workers, as well as by daily litter patrol of the site to collect any waste that has been scattered. Also, it is useful to minimise the area of the working face.

Daily waste Compaction:
This may help reduce vector and vermin hazards.

Daily waste covering:
Use of cells and interim covers can reduce fly, smell and vermin nuisance. A layer of daily cover in a landfill is called a cell. Interim covers are thicker layers of daily cover material applied to areas of the landfill that will not be worked for some time.

Compensation and resettlement:
These can be done to minimise social impacts on the displaced families.

Targeting low populated areas:
Ideally, a landfill site should have sparse population so that there may be no need for displacement and resettling. Proper town / city planning should ensure this.

Landscaping:
Both excavated sites and the closed landfill can be landscaped to make the site aesthetically acceptable. Some of the cover material and capping material can be selected so that the beauty aspect is catered for. The cover material can be derived from raised grounds so that some landscaping of the source of cover material is also simultaneously accomplished.

Safe scavenging:

The waste can be deposited in some temporary site (e.g. transfer station) to allow scavengers to pick their waste before finally being transferred to the landfill. Use of convenience transfer station for such a process can be pursued. Only organised and controlled scavenging should be permitted, and can alternatively be confined to the sloping forward surface of the working face. Potable containers must be provided to contain the salvaged materials.

Dust control:
Where possible, dusty roads should be sprayed with water. Alternatively, a dust screen in form of tree barriers may be planted along the side of the road leading to and from the site.

Risk Management and assessment (RMA)
The Purpose of RMA is to identify the environmental impacts of different situations to estimate the likelihood of risk, the magnitude of the impacts, and to identify how each risk can be managed.

Incinerators
It is not cost effective by just any institution to construct incinerators. Incineration of municipal solid waste is rarely economically feasible for developing countries. Burning the wet waste found in Africa often requires adding supplemental fuel. Furthermore, the composition of the waste often varies a great deal between neighborhoods, which make consistent and optimal operation difficult to achieve. Without proper controls, incinerators can be highly polluting, generating dioxins and depositing toxic heavy metals into water bodies. The proprietary technologies involved require

very large capital investments and have high maintenance costs.

CHAPTER 16: MANAGING SPECIAL WASTES

These are wastes requiring Special Attention. Certain wastes merit special handling and disposal because of their dangers or volume. The best option is to minimize or eliminate the generation of these wastes by encouraging users to apply cleaner production approaches and substitute materials or change processes. Those that are generated should be collected and disposed of separately from one another and away from the rest of the solid waste stream.

Hazardous waste. Wastes pose a wide range of risks. They may be chronically and acutely toxic, cause cancer, trigger birth defects, explode, corrode many materials, and cut, puncture, crush, burn and infect people and animals. Hazardous wastes endanger many different classes of people, placing waste producers, collectors, landfill workers, waste pickers, and nearby residents at risk. The leachate from a landfill may be dangerous as well; its level of toxicity is directly related to the quantity and toxicity of hazardous materials mixed in with other solid waste.

Management of hazardous wastes needs urgent attention in Africa. The variety and classes of materials and sources—from households to industrial and medical facilities—makes this particularly challenging. Action is constrained by limited financial resources to deal with these problems and ignorance or unwillingness to acknowledge the risks.

Sound management of hazardous materials includes four elements: waste reduction, segregation, safe handling, and disposal. The best solution is to not generate this waste in the first place. When this is not possible, every effort should be made to minimize generation, and generated wastes should be handled cautiously to reduce risks. Producers of hazardous waste should segregate different types of materials to make recycling easier and prevent chemical reactions or explosions.

Hazardous wastes have been defined as wastes or combination of wastes that pose a substantial present or potential hazard to humans or other living organisms because (i) such wastes are non degradable or persistent in nature (ii) they can be biologically magnified (iii) they can be lethal, or (iv) they may otherwise cause or tend to cause detrimental cumulative effects. Properties of waste material that have been used to assess whether a waste is hazardous are related to questions of safety and health. These are outlined in the table below (Tchobanoglous G et al. (1993)). Many of the products used around the home every day such as household cleaners, personal products, automotive products, paint products and garden products are toxic and can be hazardous to the environment. These products, as shown in figure 4-5, are corrosive, flammable, irritants, and poisonous.

Commercial sources of Hazardous wastes
Commercial establishments are often identified as small quantity generators of hazardous wastes. These are primarily linked to service provided such as inks from print shops, solvents from dry cleaning establishments, cleaning solvents from auto-repair shops and paints and thinners from painting contractors. In view of the fact that most commercial establishments are often required to have a pre-treatment of their

wastes prior to disposal at landfills, the major concern arises from household hazardous wastes. Special focus is therefore required for HHW. Typically, amounts of hazardous wastes found in MSW vary from 0.01-1%, with an average of 0.1% (Tchobanoglous G et al. (1993)). Of this, approximately 75-85% come from residential sources (HHW).

Table 7: Typical Characteristics of Hazardous wastes from Residential sources - Safety issues in household, commercial and industrial wastes (Adapted from Tchobanoglous G et al. (1993).

PROPERTY	Household (HHW)	Commercial & industrial
Safety-related properties		
Corrosivity	Abrasive scouring powders, ammonia and related cleaners, drain openers, oven cleaners, toilet bowl cleaner, batteries, pool acids and chlorine,	
Explosivity	Gas cylinders, pressure cookers	Gas cylinders, pressure cookers
Flammability & Ignitability	Aerosols, Furniture polish, Shoe polish, Silver polish, Spot remover, Upholstery and carpet cleaner, Nail polish remover, brake & transmission fluid, pesticides	Barium, Cadmium, Toluene, Benzene,
Health-related properties		
Irritant (allergic response)	Glass cleaners,	Tetrachloroethene, arsenic (dermatitis).
Toxicity	Household cleaners, houseplant insecticide, chemical fertilizers, automotive products e.g. antifreeze, Rubbing alcohol, medicated shampoos, personal care products e.g. hair-waving lotions	
Carcinogen		Arsenic, Cadmium, Chromium, benzene, Chloroethene, Dichloromethane, halogenated

		pesticides
Mutagen		Arsenic, halogenated pesticides
Neurotoxin		Lead

Significance of Hazardous wastes in MSW

The small amounts of hazardous wastes found in MSW are of significance because of their occurrence in all solid waste management facilities and their persistence when discharged to the environment. They may exist in solids, liquid, gaseous and semi-solid forms derived from the HHW. These influence the recovery of materials, conversion products (e.g. compost), combustion products and landfills. The HHW is largely trace organic constituents that can be separated mechanically from commingled MSW. Trace amounts have also been found in compost produced from MSW, rendering the products unusable. Toxic heavy metals such as barium, cadmium, chromium, lead, mercury and silver are especially troublesome, and they have been found in trace levels in gaseous emissions and residual materials resulting from the combustion of solid wastes. Similarly, trace organic constituents have been found in the atmosphere near landfills, in extracted landfill gas and in landfill Leachate.

These have two basic sources: (i) derived from hazardous waste themselves or (ii) produced by chemical and biological conversion reactions within the landfill. Most of these products are of concern because of long-term persistence, and pose acute to chronic toxicity problems. They produce other products by volatilisation, simple substitution, dehydrogenation (hydrolysis), Auto-oxidisation and reduction reactions. Most of these products are used, stored, and often disposed of improperly by at the household storage from where it gets mixed with the rest of the MSW. This makes further processes difficult. Therefore management of HHW at household level would form the single most important intervention point of management of the HHW, and in totality, the entire MSW management. Source separation by households is being encouraged to eliminate these constituents from solid waste processing operations.

Proposal For Household Hazardous Waste (HHW) Collection Programmes

To minimise improper disposal of HHW, product exchange programs, special collection days and permanent collection sites have been established by a number of communities.

Product exchange programs

Because paint products form a major portion of HHW, paint exchange programmes are being used in a number of communities to reduce the cost of HHW disposal. The reuse of latex-based paints has proven most successful, with up to 50% recovery. Unrecoverable paint must be either combusted in a hazardous waste combustor or disposed of in a hazardous waste landfill.

Special collection days

One of the most common approaches to HHW management is to hold one or more community waste collection days. On these days, the community members are asked to bring their HHW, at a little or no charge, to a specified location for recycling, treatment or disposal by professional waste handlers. In larger communities, several locations are used on successive days. Adequate promotion and education are essential for success of such programs. Records of 5-10% of such HHW are collected through such programs in Europe (Tchobanoglous et al, 1993) and it would be useful if even half of that is collected in Maun, Botswana.

Permanent collection sites

To increase the convenience of HHW collection programs and therefore increase participation, more and more communities are establishing permanent collection sites (e.g. fire stations, landfills city and corporation yards etc) programs involving permanent collection facilities allow citizens to drop off wastes at their own convenience. For this reason, permanent collection sites have proven to be more effective for collecting HHW than the one-day collection programs.

Suggested best practices for accomplishing these goals in the developing world include:
• Providing technical assistance and training to educate decision-makers, system operators, and the public. These efforts should strengthen stakeholders' capacity to identify cost-effective waste reduction measures, and to help design and to put in place practical hazardous waste management plans.
•   Establish   incentives,   disincentives,   or regulations to promote waste reduction where it is not otherwise cost-effective.
•   Establish   dedicated   hazardous   waste recycling and disposal facilities. Few countries in Africa operate hazardous waste treatment and disposal facilities. Thus, much of the hazardous waste generated continues to be disposed of in dumps and landfills without any provisions for segregation, containment or treatment.

• Develop systems to ensure that waste is not illegally dumped. One model that provides checks on illegal dumping is the hazardous waste manifest system in the United States, where a "paper trail" (a sequence of required documents) is generated to prove that the material reached its intended final destination.
• Explore options for contracting private sector firms that specialize in the handling and disposal of hazardous wastes.

Medical waste.
Wastes from health posts, clinics, hospitals, and other medical facilities pose serious and urgent problems in the Africa region.
These wastes can contain highly infectious organisms, sharp objects, hazardous pharmaceuticals and chemicals, and even radioactive materials. Since the various forms of healthcare waste require different types of treatment, they should be segregated at the source. General waste should be segregated from hazardous material to reduce volume: sharps should be placed in puncture-proof containers, infectious waste separated for sterilization, and hazardous chemicals and pharmaceuticals segregated into separate bins. Unfortunately, all of the available disposal options are imperfect. The most immediate threat comes from highly infectious waste. On-site treatment is generally preferred to reduce the risk of disease transmission to waste handlers, wastepickers and others. Suggested mitigation measures include:
• In rural areas, burn infectious waste in a single-chamber incinerator, if possible. This kills >99 percent of the organisms and is the best option for minimal facilities.

• In urban areas, burning is not advisable, as the fly ash, toxic gases and acidic gases pose a much greater health threat in more densely populated urban environments than in rural areas. Thus larger facilities should autoclave infectious waste. While high-temperature incineration is theoretically the best option in urban environments, in practice the equipment is rarely operated properly and disposal is highly polluting.

• In some large cities, off-site wet thermal, microwave or chemical treatment options may be available.

• The least expensive option is land disposal. If waste is to be disposed of in a dump or landfill, it should be packaged to minimize exposure, placed in a hollow dug below the working face of the landfill, and immediately covered with 2 m of mature landfill waste. Alternatively, it may be placed in a 2 m deep pit and covered in the same manner. Waste-picking must then be prevented.

Healthcare Waste (HCW)

Hazardous Clinical Wastes (HCW)

The following are the observed categories of HCW generated

1.   General Waste (Non-toxic) e.g. food remains, containers

2.   Pathological waste: Solid body parts with cancerous growth e.g. breasts

3.   Pharmaceutical waste: Dirty or expired drugs

4.   Radioactive waste (from radiology and X-ray laboratory)

5.   Wastes rich in heavy metals e.g. Thermometers (mercury), Lead acid batteries (Lead);

6.   Combustible waste e.g. cotton wools

impregnated with methylated spirits

7.   Explosive-prone wastes e.g. old unused gas cylinders;

8.   Sharps e.g. needles, scalpel blades, broken drug bottles;

9.   Infective waste e.g. blood-stained clothes and equipment;

HCW is the total waste stream generated by nursing homes, hospitals, mobile surgeries, dental surgeries, health care establishments, research facilities and laboratories. Other names are clinical waste, Medical waste, and hospital waste. Ali (2003), states that 75-90% of health care waste can be classified as non-clinical or general waste, which present no higher risk to the community than municipal waste. The remaining portion of 10-25% can be classified as clinical waste, and it is this portion that can be hazardous. This clinical waste comprises pharmaceutical, pathological, pressurized containers, wastes with high heavy metal contents, sharps, radioactive wastes, chemical wastes, infectious wastes, and Genotoxic wastes

Tires, oil, and batteries.

These three common automotive wastes cause difficulties throughout the African continent:

Stockpiled tires can spontaneously combust, producing prolonged, polluting fires. Reuse or retreading are the best alternatives available for reducing tire waste in developing and industrializing countries.

Used motor oil from auto shops is often burned as fuel, contributing to air pollution. Re-refining this oil is the best alternative, but this alternative is neither readily available nor commercially feasible in most of Africa. Lead acid batteries should not be placed in landfills— the lead is toxic, the acid corrosive and contaminated. Lead acid batteries are often recycled in small-scale foundries that are highly polluting and located in residential areas. Recycling in large facilities that have emission and environmental controls is preferable, if this option is available.

Construction and demolition debris.
Its important to prevent disposal of construction and demolition debris in dumps or landfills, as this will greatly reduce the life of the facility. Residual lead paint, mercury switches, asbestos and PCBs can also make this debris toxic. Arrange for the return of unused construction materials, recovery of all reusable or recyclable materials, and on-site separation of different waste materials to simplify reuse. The UN Environment Programme's International Sourcebook on Environmentally Sound Technology for Municipal Solid Waste Management recommends the following best practices for construction and demolition debris:
• Inventory control and allowance for return of construction material. This ensures that unused materials will not be disposed of unnecessarily.
• Selective demolition. This involves dismantling, often for recovery, selected parts of buildings to be demolished before the wrecking process is initiated.
• On-site separation systems. Use multiple smaller containers instead of a single roll-off or compactor.
• Crushing, milling, and reusing secondary stone and concrete materials. There can be a tie-in to approved road construction material specifications.

CHAPTER 17: CASE STUDY

The current practices for managing the healthcare waste in Maun, Botswana

This chapter presents a case study of Maun city, in the North west of Botswana, Southern Africa. It borrows from a study undertaken in the city in 2002/2004 period.

Most HCW is incinerated at the central Government incinerator based at Maun general Hospital. There is no waste separation – so most of the waste is incinerated. A little of the other general waste from kitchen, together with the incineration residue, are taken to the public disposal site by council Trucks. Some of the HCW is disabled by thermal treatment (oven and autoclave) or chemical disinfection (using bleaching powder, detergents such as OMO etc). However, there are some key problems associated with the current practices of managing HCW in Maun. These are discussed in the section below.

The key problems with the current practices of managing clinical wastes in Maun, Botswana
There are serious concerns about hazardous waste management in Maun. These concerns include:
- Produces highly concentrated solid residues such as ash (with high leaching potential)
- Air pollution/ global warming from the combustion gases. The combustion gases such as CO_2, SO_x, NO_x etc are vital greenhouse gases, which, if allowed to accumulate in the atmosphere, can cause serious climatic catastrophe in the future.

- Capital-intensive incineration method is used; this is expensive to establish.
- The incineration ends up wasting energy which could be recovered from the energy-rich combustible wastes;
- The gases produced (e.g. SO_2, SO_3, NOx) could cause respiratory diseases to human beings working or residing around;
- The gases produced (e.g. SO_2, SO_3, NOx) could cause wearing of metallic (especially iron and steel based) surfaces by rusting, wear and tear, as they would produce weak acids with atmospheric water vapor. This weak acid would be a catalyst to the wear and tear processes, and would reduce the lifespan of a number of structures such as roofs.
- Denies human scavenger a chance of recovering useful material. In Botswana, resource use efficiency is very low due to the misguided attitude that the government has money. There is therefore high rate of misuse of supplies, with very high reuse potential (at times usable as some is disposed of even before use) and could be recovered if scavengers were given a chance, albeit with some health precautions.

Proposals on how to improve the current hazardous clinical waste management
The current practices can be improved by:
(i) Installing a wet scrubber to take up the potential air pollutants, especially acidic ones such as SOx, NOx etc.
(ii) Installing a long chimney to help disperse the gases; this, however, is another method of pollution as the hazardous gases are only expelled from the point of production but transferred elsewhere (dilute and attenuate

related scenario)

(iii)   Conduct regular health surveillance on the people residing and working around the site. This would help monitor any significant changes in their health that would be attributed to the exhaust gases.

(iv)   Possibility of energy recovery from the system could be investigated, and if viable, installed.

NB: SO_x and NO_x are used here to mean any oxides of Sulphur and Nitrogen respectively. Since these 2 elements have different oxidation states, they have the ability to produce different kinds of oxides depending on the prevailing conditions, especially, temperature, pressure and Oxygen supply in the system of their origin.

CHAPTER 18: ENVIRONMENTAL ASSESSMENT AS A TOOL FOR DECISION MAKING IN ENVIRONMENTAL MANAGEMENT

General introduction

An Environmental Impact Assessment (EIA) is a study of the effects of a proposed action on the environment; it is a critical examination of the positive and negative effects of a project on the environment. In this regard the environment includes all relevant aspects of the natural and human resources. The EIA evaluates the expected effects on human health, the natural environment (air, soil, land, water, flora and fauna) and on property. It is a preventive process that seeks to minimise the adverse effects of a project on the environment. The study is therefore normally given a multi-disciplinary approach. Its goal is to ensure that decisions on proposed projects and activities are environmentally friendly and sustainable. It should be done very early at the feasibility stage of a project. In other words a project should be assessed for its environmental feasibility. The EIA compares various alternatives by which the project could be realized and seeks to identify the one that represents the best combination of economic and environmental costs and benefits. Alternatives include variation in location, methods (process technology and construction), and size.

EIA is based on predictions. It attempts to predict the changes in environmental quality, which would result from the proposed project/action. The EIA attempts to weigh environmental effects on a common basis with economic costs and benefits and finally it is a decision making tool. The EIA is a procedure used to examine the environmental consequences, both beneficial and adverse, of a proposed development project and to ensure that these effects are taken into account in project design. EIA should be viewed as an integral part of the project planning process, and done at the initial stages of project development. It is therefore ideal to conduct it as part of the project development process as a decision-making tool which later guides decisions on abandonment, modification or implementation of a project.

The EIA Objectives, Procedures And Guidelines

The production of goods and services to meet global population demands has occasioned a number of activities that have depleted the globe's natural resources and in several instances contributed to environmental degradation through pollution. These activities done in the pursuit of economic development have also caused the loss of several species of plants and animals and now threaten the existence of man himself, if left uncontrolled. Recognition of the question of the globe's capacity to sustain these activities and the general environmental problems associated with them, which are common at the community, national, regional and international levels, led to a number of international conferences (starting at Stockholm in 1972), treaties, conventions, and protocols on the management of the earth's resources in an effort to ensure sustainable economic development. In 1987 the United Nations Environment Programme (UNEP) adopted a set of goals and principles on EIA. At the national level legislation has been enacted in almost every country. This applies to

Kenya as well. The goal of an EIA is to ensure that decisions on proposed projects and activities are environmentally sustainable. The objectives of the EIA are:

i. Identify likely impacts through screening and scoping;
ii. Outline relevant legislation with respect to the proposed project;
iii. Generate baseline data for monitoring and evaluation of impacts Description of the Environment
iv. Description of the Proposed Project
v. Predict likely changes on the environment as a result of the development
vi. Identify Significant Environmental Impacts
vii. Identification and Analysis of Alternatives, including evaluating the impacts of the various alternatives of the project, and the no-project scenario.
viii. Propose Mitigation Action and measures / Mitigation Management Plan for the significant negative adverse impacts on the environment
ix. Monitoring and evaluation Programme

Methodology Outline
The general steps followed in a typical Environmental assessment are:

i. Environmental screening upon which a full impact assessment was found necessary,
ii. Environmental scoping during the preliminary site visit,
iii. Consultation and public participation (On a continuing basis),
iv. Assessment of anticipated impacts,
v. Impact mitigation
vi. Report preparation and discussions.

Environmental Screening

This step is always applied to determine whether environmental impact assessment is required and to know what level of environmental assessment was necessary as per the gazetted EIA guidelines. According to the 2nd schedule of EMCA (1999), Construction of an earth-moving project such as a landfill, is covered for EIA under the general category (1) which covers: An activity out of character with its surrounding; (ii) A structure of a scale not in keeping with its surrounding; and (iii) major changes in land use. Such a project is out of character (since it does not exist in many cities in the developing countries such as Kenya), is of large scale, and is likely to cause major changes in land use. These straightaway qualify the an earth-moving project such as a landfill, a very detailed EIA prior to implementation. This is largely to satisfy the construction and operation, drilling requirement and for identification of environmental monitoring tools.

Environmental Scoping

The scoping process need to be carried out in order to narrow down onto the most critical issues requiring attention during the assessment. The exercise may involve categorizing the environment into physical, natural/ecological, social, economic and cultural aspects. Some level of environmental impacts assessment likely to be found necessary after pre-assessment of the following issues.

- Impacts on Physical Environment
- Impacts on Natural/Ecological Environment

- Impacts on Social, Economic and Cultural Environment

Environmental Impact Assessment Guidelines
The EIA guidelines require that EIA be conducted in accordance with the issues and general guidelines spelt out in the second and third schedules of the regulations. This includes coverage of the issues on schedule 2 (ecological, social, landscape, land use and water considerations) and general guidelines on schedule 3 (impacts and their sources, project details, national legislation, mitigation measures, a management plan and environmental auditing schedules and procedures).

Relevant Policy, Legislative And Regulatory Framework for an EIA in Kenya

General Overview
This section describes the policy and legal basis within which a typical landfill project may be implemented. Regulations and standards applicable to the project are referred to. EIA is a tool for environmental conservation and has been identified as a key component in new project implementation. At the national level, Kenya has put in place necessary legislation that requires EIA to be carried out on every new major project, activity or programme (EMCA, 1999), and that a report be submitted to the National Environmental Management Authority (NEMA) for approval and issuance of relevant certificates and/or licenses. To facilitate this process, regulations on EIA and environmental audits (EA) have been established under the Kenya Gazette Supplement No. 56 of 13th June

2003. Some aspects have since been modified accordingly.

Besides, a number of other national policies and legal statutes have been reviewed to enhance environmental sustainability in national development projects across all sectors. Some of the policy and legal provisions are normally presented as shown in the example below, with some sub-sections.

Environmental Policies
A sanitary landfill involves excavation, lining and deposition of potentially polluting wastes in the completed facility. Even though there is lining, with lots of controls and monitoring, ground water is at risk any time a landfill is developed. This is due to leachate pollution of the water. As such, all legal provisions of water, among others, apply.

The National Policy on Water Resources Management and Development
Kenya is water scarce with per capita availability of only 647 m3 per year. There is Low-level investment in the water sector at the moment. The Water Sector Reforms and Regional Co-operation are geared towards improving the situation. While the National Policy on Water Resources Management and Development (1999) enhances a systematic development of water facilities in all sectors for the promotion of the country's socio-economic progress, it also recognizes the by-products of these processes as water. It, therefore, calls for the development of appropriate sanitation systems to protect people's health and water resources from institutional pollution. Industrial and development activities, therefore, should be

accompanied by corresponding waste management systems to handle the wastewater and other wastes emanating from there. The same policy requires that such projects should also undergo comprehensive EIAs that provide suitable measures to be taken to ensure environmental resources and people's health in the immediate neighbourhood and further downstream are not negatively impacted by any proposed projects. As a follow-up to this, EMCA, 1999 requires annual environmental audits to ensure continuous improvements.

In addition, the policy provides for charging levies on wastewater (leachate) based on quantity and quality (similar to polluter-pays-principle) of effluent. Further, the policy requires those contaminating water to meet the appropriate cost on remediation, though the necessary mechanisms for the implementation of this principle have not been fully established under the relevant acts. However, the policy provides for establishment of standards to protect water bodies receiving wastewater, a process that is ongoing.

Policy Guidelines on Environment and Development

Among the key objectives of the Policy Paper on Environment and Development (Sessional Paper No. 6 of 1999) are to:
- Ensure that all development policies, programmes and projects mainstream the environment.
- Ensure that independent EIA reports are prepared for all development initiatives before implementation,

- Come up with effluent treatment standards that will conform to acceptable health guidelines.

Legal Aspects of EIA

Application of national statutes and regulations on environmental conservation suggest that developers have a legal duty and responsibility to discharge wastes of acceptable quality to the receiving environment without compromising public health and safety, and any related biodiversity. This position enhances the importance of an environmental audit for the operations at the site to provide a benchmark for its sustainable operation. The key national laws that govern the management of environmental resources in the country have been discussed in EIA reports and wherever any of the laws contradict each other, the Environmental Management and Co-ordination Act 1999 prevails.

Water Act 2002

This act established the Water Resources Management Authority, the Catchment Area Management Committees; the Water Services Trust Fund; the Water Appeal Board; the Water Services Regulatory Board and seven Water Services Boards. Repealed the Water Act Cap.372 and certain provisions of the Local Government Act, thereby allowing Stakeholder and Community participation in WRM&D, the formation of Water Resources Users Associations, and Water Services Providers Part II, section 18, of the Water Act, 2002 provides for national monitoring and information systems on water resources. Following on this, sub-section 3 allows the Water Resources Management Authority to demand from any person or institution, specified information, documents, samples or materials on water resources. Under these rules, specific records may require to be kept by a facility operator and the information thereof furnished to the authority. Section 73 of the Act allows a person with license (licensee) to supply water and make regulations for purposes of protecting against degradation of water sources.

The Public Health Act (Cap. 242)
Part IX, section 115, of the Act states that no person/institution shall cause nuisance or condition liable to be injurious or dangerous to human health. Section 116 requires Local Authorities to take all lawful, necessary and reasonably practicable measures to maintain their jurisdiction clean and sanitary to prevent occurrence of nuisance or condition liable to be injurious or dangerous to human health. Such nuisance or conditions are defined under section 118 as waste pipes, sewers, drains or refuse pits in such a state, situated or constructed as in the opinion of the medical officer of health to be offensive or injurious to health. Any noxious matter or waste water flowing or discharged from any premises into a public street or into the gutter or side channel or watercourse, irrigation channel or bed not approved for discharge is also deemed as a nuisance. Other nuisances are accumulation of materials or refuse which in the opinion of the medical office of health is likely to harbour.

On the responsibility of local authorities, Part XI, section 129, of the Act states in part "it shall be the duty of every local authority to take all lawful, necessary and reasonably practicable measures for preventing any pollution dangerous to health of any supply of water which the public within its district has a right to use for drinking or domestic purposes....". Section 130 provides for making and imposing regulations by the local authorities and others the duty of enforcing rules in respect to prohibiting use of water supply or erection of structures draining filth or noxious matter into water supply as mentioned in section 129. Part XII, Section 136, states that all collections of water, sewerage, rubbish, refuse and other fluids which permits or facilitates the breeding and multiplication of pests shall be deemed nuisances and are liable to be dealt with in the manner provided by this Act.

The Environment Management and Co-ordination Act (EMCA), 1999
Part II of the Environment Management & Co-ordination Act, 1999 states that every person in Kenya is entitled to a clean and healthy environment and has the duty to safeguard and enhance the environment. In order to ensure that this is achieved part VI, section 58, of the same Act directs that any proponent of a new project should carry out an EIA and prepare an appropriate report for submission to the authority (NEMA), who in turn issues a license as appropriate. The second schedule of the same Act lists proposed urban development activities as among the facilities that should undergo environmental impact assessments. Part VIII, section 72, of the Act prohibits discharging or applying poisonous, toxic, noxious or obstructing matter, radioactive or any other pollutants into aquatic environment. Section 73 requires that operators of projects, which discharge effluent or other pollutants,

submit to NEMA accurate information about the quantity and quality of the effluent. Section 74 demands that effluent generated from any trade undertaking are discharged only into the existing sewerage system upon issuance of a license from the Authority.

Consultation and public participation in environmental impact assessment
The importance of public participation in decision-making in environmental matters is further highlighted by the requirement for environmental impact assessment study report under Part VI of the Act. Any person, being a proponent of a project is required to apply for and obtain an E.I.A licence from NEMA before he can finance, commence, proceed with, carry out, execute, or conduct any undertaking specified in the 2nd Schedule of the Act. The EIA study report is published for two successive weeks in the Gazette and in a newspaper circulating in the area or proposed area of the project and the public is given a maximum period of sixty days for inspection of the report and submission of oral or written comments on the same. Any person may extend this period on application. The EIA process, thus, gives individuals and communities a voice in issues that may bear directly on their health and welfare and entitlement to a clean and healthy environment.

Offenses related to environmental assessments and inspections
The following profiles the environmental offences as recognized by EMCA (1999) in Kenya
1d: offences as recognized by EMCA act 1999:
137 – Offences relating to inspection.

138 – Offences relating to environmental impact assessment.
139 – Offences relating to records.
140 – Offences relating to standards.
141 – Offences relating to hazardous wastes, materials, chemicals and radioactive substances.
142 – Offences relating to pollution.
143 – Offences relating to environmental restoration orders, easements, and conservation orders.
144 – General Penalty.
145 – Offences by bodies corporate, partnership, principals and employees.
146 – Forfeiture, cancellation and other orders.
147 – Regulations.
148 – Existing laws relative to the environment.

Environmental offences in EMCA
These are highlighted in Section 137, as follows:

(a) hinders or obstructs an environmental inspector in the exercise of his duties under this Act or regulations made thereunder;
(b) fails to comply with a lawful order or requirement made by an environmental inspector in accordance with this Act or regulations made thereunder;
(c) refuses an environmental inspector entry upon any land or into any premises, vessel or motor vehicle which he is empowered to enter under this Act or regulations made thereunder;
(d) impersonates an environmental inspector;
(e) refuses an environmental inspector access to records or documents kept pursuant to the provisions of this Act or regulations made thereunder;
(f) fails to state or wrongly states his name or address to an environmental inspector in the

cause of his duties under this Act or regulations made thereunder;

(g) misleads or gives wrongful information to an environmental inspector under this Act or regulations made thereunder;

(h) fails, neglects or refuses to carry out an improvement order issued under this Act by an environmental inspector;

commits an offence and shall, on conviction be liable to imprisonment for a term not exceeding twenty four months, or to a fine of not more than five hundred thousand shillings, or both.

These are also highlighted in Section 1387, as follows:

(a) fails to submit a project report contrary to the requirements of section 58 of this Act;

(b) fails to prepare an environmental impact assessment report in accordance with the requirements of this Act or regulations made thereunder;

(c) fraudulently makes false statements in an environmental impact assessment report submitted under this Act or regulations made thereunder; commits an offence and is liable on conviction to imprisonment for a term not exceeding twenty four months or to a fine of not more than two million shillings or to both such imprisonment and fine.

The Role Of Environmental Audit And Monitoring

Part 7 of the Act (Sections 68-69) gives NEMA the responsibility of carrying out environmental audits of all activities that are likely to have significant effect on the environment. In consultation with lead agencies, the Act also authorises NEMA to carry out environmental monitoring of all environmental phenomena and operations of industry, projects or activities to determine their impacts.

When to conduct EIA:
EIA is considered part of the project development process, and is thus done at the initial stages of project development. It is a decision making tool and should guide whether a project should be implemented, abandoned or modified prior to implementation.

Issues considered in the EIA as per the EMCA 1999(NEMA (2004, 2006, 2005, 2003a and 2003b):
i.   Ecological considerations, including biodiversity, sustainable use, and ecosystem maintenance;
ii.   Social considerations, including: economic impacts; social cohesion or disruption; effects on human health; immigration or emigration; communication; and effects of culture or objects of cultural value;
iii. Landscape issues, including views opened up or closed, visual impacts; compatibility with surrounding area; and amenity opened up or closed;
iv.   Land use, including: effects on current land uses and land use potentials in the project area;
v.   Effects of proposal on surrounding land uses and land use potentials; and possibility of multiple uses.
vi.   Water: water resources (quality and quantity of sources; and drainage patterns or systems.

The EIA process:

The following steps are undertaken in a typical environmental assessment in Kenya:

a) Development and submission of a project report for projects or activities which are not likely to have significant environmental impacts or those for which EIA study is required (GoK (1999 and GoK (1994). . Then the EIA process, if done, is as follows (NEMA (2004, 2006, 2005, 2003a and 2003b);

b) Scoping and drawing up terms of reference (TOR) for the study for approval by the authority;

c) Gathering baseline information through investigation, research and subsequent submission of EIA report to the authority. At submission, ten copies of the report, a copy of the report in Compact Disc , and a banker's cheque for 0.1% of the project cost payable to NEMA. The reports are thereafter distributed to NEMA review experts for peer review, normally within 21 days of project submission.

d) Review of EIA study report by the authority and NEMA recommended lead agencies.

e) Decision on the EIA study report, which may include non-conditional approval, conditional approval, or rejection.

f) If approved, then the significant impacts and their proposed mitigation measures are prepared by NEMA and sent to the proponent to advertise in the media for further stakeholder input. A maximum of 90 days are allowed to pass during which these views are submitted to NEMA. In the absence of a negative concern from stakeholders within the 90 days, an EIA certificate is issued to the proponent.

g) Implementation of the project / OR Appeals: The latter applies if the project is not approved.
h) Monitoring the project (Based on the indicators identified at the EIA)
i) Auditing the project (Environmental audit).

Impact Identification

The impacts of a project may fall in the following categories:

a. Impacts on Physical Environment

- Water quality,
- Soil quality and land,
- Air quality,
- Land issues with respect to being out of character with the surrounding.

b. Impacts on Natural/Ecological Environment

- Vegetation typical of the area. Field observations provided the main baseline situation,
- Wildlife with respect to displacement of species, migration of species, enhanced breeding of pests, snakes etc.,
- Effects on air quality;
- Effects on soil quality;
- Area topography effects, etc.

c. Impacts on Social, Economic and Cultural Environment

- Provision of employment, learning opportunities, saving of foreign exchange,
- Settlement patterns or disturbance to the public,
- Changes in land-use,
- Income generation opportunities,

- Introduction of nuisances and risks associated with student riots and demonstrations.
- Community services;
- Cultural exchange and enrichment

Basic Checklist Used To Compile The Description Of The Environmental Setting
This checklist lists some factors, which should be considered in describing the environment. This description of the environmental setting is a record of conditions prior to implementation of the proposed project. It is primarily a benchmark against which to measure environmental changes and to assess impacts.

1. Basic Land Conditions
   a. Geological Conditions
   - Major land formations (valleys, rivers)
   - Geologic structures (sub-strata, etc.)
   - Geologic resources (minerals, oil, etc.)
   - Seismic hazards (faults, liquefaction, tidal wave etc.)
   - Slope stability and landslide potential

   b. Soil Conditions
   - Soil conservation service, classification
   - Hazard potential (erosion, subsidence or expansiveness)
   - Natural drainage rate
   - Sub-soil permeability
   - Run-off rate
   - Effective depth (inches)
   - Inherent fertility
   - Suitability for method of sewage disposal

   c. Archaeological value of site

2. Biotic Community Conditions

a. Plant

- General type and dominant species
- Densities and distributions
- Animal habitat value
- Historically important specimen
- Watershed value
- Man-introduced species
- Endangered species (location, distribution and conditions)
- Fire potential (chaparral, grass, etc.)
- Timber value
- Specimen of scientific or aesthetic interest

b. Animal

- General types/dominant species (mammal, fish, fowl, etc.)
- Densities and distribution
- Habitat (general)
- Migratory species
- Game species
- Man-introduced species (exotic species)
- Endangered species
- Commercially valued species

3. Watershed Conditions

- Water quality (ground water and surface water)
- Source of public or private water supply on-site
- Watershed importance (on-site and surrounding area)
- Flood plain importance (on-site and surrounding area)
- Water run-off rate

- Streamside conditions (habitat conditions and stream flow rate)
- Location of wells, springs
- Marshlands, lakes, ocean frontage importance

4. Airshed Conditions
- General climatic type
- Air quality
- Airshed Importance
- Wind hazard area (min/max speeds)
- Odour levels
- Noise levels
- Rainfall (average)
- Temperature (average highs and lows)
- Prevailing winds (direction and intensity)
- Fog conditions (hazard potential)

Question:
Using a case study of an environmental impact assess met of your choice, reflect on the decision-making process using analytical strategic environmental assessment framework as a reference methodology. Your discussion must be no longer than 800 words) (15% of marks)- i.e. (7.5 of total module marks.)2

The Analytical Strategic Environmental Assessment (ASEA)
Analytical Strategic Environmental Assessment (ASEA) is a modified and a more systematic form of the formal Environmental Impact Assessment (EIA), which strives to improve the decision-making aspect of the environmental assessment process (Dijkstra (2003), Lee and George (2000) and Canter (1977). EIA creates

an impression that decision-making is at only one point / instant (i.e., instantaneous), and involves only one person / party, the regulatory authority. In real sense, however, environmental assessment involves many decisions, made by different people/parties, and at different stages. Some of these decision makers assumed by the EIA, and areas where they make decisions in the project cycle are:

a) The financier: can decide to fund or not to fund the project at any stage even after initial financial approval. An example is the Sondu-Miriu project in Kenya, which was recommended by almost all internal authorities, but stopped by the civil society at implementation stage. This created a chain reaction- where the financier stopped the process; the then central government also lost interest, etc.

b) The developer: may also change mind at any stage, thereby stopping all other subsequent stages. This may result from change of policy, priorities or governments.

c) The civil society: This has a very strong mobilizing power both locally and internationally, and can stop even projects approved by the highest authority in the land, or give the project a very bad publicity. An example is the case in Botswana Bushmen relocation from their customary homes. This was brought to the international limelight by an international NGO. Many residents refused to relocate (especially after the publicity) despite threats of forceful eviction, and only moved after essential facilities and services were completely withdrawn. Another example is the case of KEL chemicals in Thika, (Kenya) in which a priest mobilized faithfuls to demonstrate against pollution from the fertilizer and sulphuric acid manufacturing industry.

d) The politicians: can decide that no such a project is developed in his / her constituency or ward, and may mobilize constituents for backing;

e) The central or devolved Government: The president or governor can give an executive directive (superior to the decision of the regulatory authority).

f) Parliament: the parliamentary committee can also decide the fate of a project.

g) Host community: Can demonstrate for or against a project on their own volition.

h) International laws on commonly shared resources e.g. international resources such as waters (e.g. Okavango delta (Botswana/Namibia); East African Lake Victoria (Kenya/Uganda/Tanzania). In the case of Lake Victoria, a law bars any East African country from developing projects involving drawing significant amounts of water from the lake- because of an existing agreement between the Colonial East Africa and Egyptian Government in 1929 (Afullo, 2003).

i) The professional body(ies): can decide the fate of a project;

j) The Landowner: Can change mind about sold or leased land deals. At times willingness to accept compensation prevails so that despite the developer's lucrative compensation offers for resettlement of the resident(s), the latter may be adamant, thereby completely installing, or delaying a project. This may change financial plans, which may render the project uneconomical.

k) The construction sector: can dictate terms on whether to support a project or not; they may sabotage a project (e.g. if tender offered unfairly) even if approved by the regulatory authority- leading to its collapse.

l) The neighboring residents or country: An example is the proposed POPA falls dam for generation of hydroelectric power by the Namibia Government. Because the proposed dam intended to use the internationally shared waters. This proposed project has been vehemently rejected by the Botswana Government citing ecological repercussions. This is a situation where cumulative environmental impact assessment ought to have been conducted.

Decision making in EIA

In general, decision-making can take one or a combination of the following forms (adapted from Dijkstra, (2003; Lee and George (2000) and Canter (1977):

a) Dictatorial: One dominant decision maker (e.g. regulatory authority) makes a unilateral decision without engaging the other stakeholders at all. This has been the most dominant system in most governments and internationally funded projects. The decision lacks credibility, and its product mostly unsustainable;

b) Participatory: Where the regulatory authority acts as a participant alongside other stakeholders in a process facilitated by a professional body. Reasoned participatory decisions are made in such gatherings by serious deliberations and consensus. After this stage, no deviations are expected from the final authority- only approving the collective decision reached by the gathering.

c) Professional approach: Where the regulatory authority delegates the decision making to an appointed body (e.g. a consultancy team), which them makes its recommendations- which are then applied as it is. The consultancy may involve the publics and stakeholder sessions. However, these gatherings are rarely sufficient. The regulatory authority comes in just as an observer during the process to avoid influencing the direction of decision to be reached by the professional body.

d) Combined / Hybrid approach: This is whereby the participatory process is reinforced with the professional approach. The suitability hybrid approach depends on the nature of the project. However, it is advantageous because sufficient technical aspects are availed (which is likely to have weight in decision making by the regulatory authority) and shared among the stakeholders. If the technical aspect is given a participatory face, then the decisions reached by hybrid approach are likely to be most appropriate.

The need for ASEA

The need for ASEA arose due to the following limitations of the decision making stage of the EIA

a) EIA is a project – based assessment whose results are not applicable to a wide range of other situations. As such, it is expensive to carry out for every proposed development project.

b) There are many social, political and regulatory constraints in the formal EIA.

c) EIA never takes into account temporal and spatial aspects of other related projects, i.e. the cumulative effects are ignored, making the project under consideration seems like an island around which nothing else happens.

d) The decision makers may take the top-bottom approach (i.e. dictatorial) so that the other stakeholders are simply informed of the decision without prior adequate involvement.

e) Delays in decision making often lead to other stages of the project progressing, leading to less, if any, incorporation of any recommendations into the rest of the project cycle;

f) Decisions are never made open to other stakeholders, leading to reduce contribution. This compromises the quality and acceptability of the final product.

When decisions are made open / public, there are often cases where no explanation or reason is provided for the decisions, some of which never seem to have taken into account the inputs from other stakeholders; (Dijkstra, 2003). ASEA therefore strives to help improve the decision-making stage(s) of the EIA by ensuring basic rules of are observed; these can be considered as the terms of reference (TOR) of the ASEA. ASEA thus operates on the following tenets (as adapted from Dijkstra T (2003); canter (1977); and Lee and George (2000)):

a) Timeliness: Decision made at the right time so that it can be used appropriately at the next stage in the project cycle

b) Transparency / accountability: Decision made in an open, systematic way, which can be easily understood by the stakeholders; the decision makers are also ready to take responsibility of the consequences of the decision made.

c) Inclusiveness / Wide participation: This means appropriate consultation and stakeholder / public involvement is ensured before any decision can be made. This gives the decision more acceptances, and the decision makers acquire some respect and credibility.

d) Credibility: An open, accountable, systematic, inclusive and transparent decision-making process imparts virtues of credibility into the process and on the decision makers. This gives them more respect and trust. The final document and project resulting thereof is also credible and more acceptable to the stakeholders.

e) Comprehensive: This means that the decision made should have taken into account all possible factors and inputs in the circumstances.

ASEA is a more systematic method which strives to combine the dual positive attributes of the formal EIA and the decision making sciences. ASEA, being a systematic procedure, goes through the following stages (as adopted from Dijkstra, 2003; lee and George, 2000; and Canter, 1977):

- Problem identification
- Gathering of information
- Public / stakeholder involvement / further generating and sharing of information
- Decision-making / implementation
- Monitoring
- Evaluation of the situation (a follow-up)

This systematized decision-making process makes it accountable, and based on the other stages of the cycle, including consideration of cumulative effects of the project. The decision making process is therefore treated as a continuum, taking place throughout. All these positive attributes render ASEA more credible, especially when combined with the strategic environmental assessment which goes beyond the narrow project-based approach to a wider and more open and sustainable policies, plans and programs. These two tools can be more applicable in the developing countries with no EIA legislation, and most projects pollute unsustainably. Incorporating the Strategic and analytical environmental aspects into third world (LDCs) development can also help reduce the cases of projects which have failed environmental standards elsewhere being imported into these countries, as the LDCs desperately seek partnership with powerful countries in the name on welcoming investors.

The Environmental Assessment practice in Kenya

INTRODUCTION:
Goal and objectives of EIA:
According to NEMA (2004, 2006, 2005, 2003a and 2003b), and GoK (1999 and GoK (1994), the goal of EIA is to ensure that decisions on proposed projects and activities are environmentally sustainable.

The objectives of EIA are:
- To identify impacts of a project on the environment;
- To predict the likely changes on the environment as a result of the development;
- To evaluate the impacts of the various alternatives on the project (including a no project alternative);

- To propose mitigation measures for the significant negative impacts of the project on the environment
- To generate baseline data for monitoring and evaluating impacts, including mitigation measures during the project cycle;
- To highlight environmental issues with a view to guiding policy makers, planners, stakeholders and government agencies to make environmentally and economically sustainable decisions.

Which projects require EIA?

The projects to be subjected to EIA are specified in the second schedule of EMCA 1999 (GoK (1999 and GoK (1994). This includes coverage of the issues on schedule 2 (ecological, social, landscape, land use and water considerations) and general guidelines on schedule 3 (impacts and their sources, project details, national legislation, mitigation measures, a management plan and environmental auditing schedules and procedures)(NEMA (2004, 2006, 2005, 2003a and 2003b),The full list is as shown below:

General: (i) an activity out of character with its surroundings; Any structure of a scale not in keeping with its surrounding; Major changes in land use. These include: Urban development, Transport, Dams, rivers, water resources; Aerial spraying; Mining, including quarrying and open cast extraction; Forestry related activities; Processing and manufacturing industries; Electrical infrastructure; Management of hydrocarbons; Waste disposal; Natural conservation areas; Nuclear reactors; And major developments in biotechnology

Issues considered in the EIA as per the EMCA 1999(NEMA (2004, 2006, 2005, 2003a and 2003b):

- Ecological considerations, including biodiversity, sustainable use, and ecosystem maintenance;
- Social considerations, including: economic impacts; social cohesion or disruption; effects on human health; immigration or emigration; communication; and effects of culture or objects of cultural value;
- Landscape issues, including views opened up or closed, visual impacts; compatibility with surrounding area; and amenity opened up or closed;
- Land use, including: effects on current land uses and land use potentials in the project area;
- Effects of proposal on surrounding land uses and land use potentials; and possibility of multiple uses.
- Water: water resources (quality and quantity of sources; and drainage patterns or systems.

The EIA process:
Development and submission of a project report for projects or activities which are not likely to have significant environmental impacts or those for which EIA study is required (GoK (1999 and GoK (1994). . Then the EIA process, if done, is as follows (NEMA (2004, 2006, 2005, 2003a and 2003b),:

- Scoping and drawing up terms of reference (TOR) for the study for approval by the authority;
- Gathering baseline information through investigation, research and subsequent submission of EIA report to the authority. At

submission, ten copies of the report, a copy of the report in Compact Disc , and a banker's cheque for 0.1% of the project cost payable to NEMA. The reports are thereafter distributed to NEMA review experts for peer review, normally within 21 days of project submission.

- Review of EIA study report by the authority and NEMA recommended lead agencies.
- Decision on the EIA study report, which may include non-conditional approval, conditional approval, or rejection.
- If approved, then the significant impacts and their proposed mitigation measures are prepared by NEMA and sent to the proponent to advertise in the media for further stakeholder input. A maximum of 90 days are allowed to pass during which these views are submitted to NEMA. In the absence of a negative concern from stakeholders within the 90 days, an EIA certificate is issued to the proponent.
- Implementation of the project / OR Appeals: The latter applies if the project is not approved.
- Monitoring the project (Based on the indicators identified at the EIA)
- Auditing the project (Environmental audit).

Who administers EIA:
NEMA is mandated by the EMCA number 8 of 1999 to administer the EIA (GoK (1999 and GoK (1994)

Who conducts EIA?
Individual experts or firms of experts registered by NEMA are the only ones mandated top do EIA / EA studies. A register of EIA experts is available in the authority's headquarters,

district provincial offices and can be accessed upon payment of a fee of Ksh 200 (USD 2.8) (NEMA (2004, 2006, 2005, 2003a and 2003b),.

Public participation in EIA / EA studies:
The law requires that during the EIA process, a proponent shall, in consultation with the authority, seek the views of persons who may be affected by the project or activity through posters, newspapers, and radio; hold at least three public meetings with the affected parties and communities. The public participates either by submitting written proposals or making oral comments. Such comments are considered in reviewing the EIA study report (NEMA (2004, 2006, 2005, 2003a and 2003b).

Who pays for EIA?
The project proponent pays for the entire EIA process. It identifies and recruits and pays the consultant, who must be a NEMA registered and licensed expert. Thereafter, the proponent, through its consultant, pays 0.1% of the project cost (or Kshs 10,000, whichever is higher) to NEMA when the reports are being submitted to NEMA (NEMA (2004, 2006, 2005, 2003a and 2003b),.

EIA Fee payable:
- Lead expert pays a registration fee of Kshs 3000 (9000 if non citizen), and an annual practicing license of Ksh 5,000 (15000 if non citizen);
- Associate expert pays a registration fee of Kshs 2000 (6000 if non citizen), and a n annual practicing license of Ksh 3,000 (9000 if non citizen);

- A firm of experts pays a registration of Ksh 5,000, and an annual practicing fee of 20,000;
- An EIA license fee of 0.1% of total project cost payable to NEMA by proponent (nothing is paid by proponent for EA);
- EIA license surrender, transfer or validation fee of Ksh 5,000

Compliance to EA:/EIA (NEMA (2004, 2006, 2005, 2003a and 2003b),
Generally, the following apply as far as compliance is concerned:
- A proponent shall not implement a project likely to have a negative environmental impact, or for which an EIA is required by the EMCA 1999 or regulations issued under it unless an EIA has been concluded and approved according to law
- No licensing authority under any law in force in Kenya shall issue a trading, commercial or development permit for any project for which an EIA is required or for a project / activity likely to have a cumulative significant negative environmental impact unless the applicant produces an EIA license issued by the authority.

Current limitations of the EIA as an Environmental legislation and process
The EIA legislation is too specific, and recommends an EIA. EIA, however, is project based, and has serious limitations shown below. Instead, a strategic environmental assessment approach could be taken. Strategic Environmental Assessment (SEA) is an improvement of the formal Environmental Impact Assessment (EIA) involving the analysis of policies, plans and programs. The EIA only

deals with project-based impacts. Thus EIA is short-lived, and addresses only issues directly relevant to the project, within its cycle from conception, up to evaluation. Secondly, the EIA mostly stops immediately after project approval by the relevant authority i.e., at decision-making stage. SEA addresses cumulative (additive and synergistic) effects of plans, policies and programs in a wider framework that would also cover project-based assessments. SEA sets the legal, institutional, economic, social and environmental framework for broader operation of EIAs ((Lee and George (2000), Dijkstra (2003) and Canter (1977)).

Due to these limitations of the EIA, SEA strives to fill some of the gaps. The SEA approach deals with impacts in a wider scale, and covers more time; it is a continuous process as the assessments are part of an institution's policies, plans and programs. Generally, projects, programs, plans and policies have varied scope, with the project being the lowest and policy is the highest. The relationship can be illustrated as follows:

A policy may cover a whole country or region, and therefore tends to have a cumulative (i.e., temporal and spatial) consideration of impacts. Project-based EIAs, on the other hand, address narrower site-based or local impacts (canter, 1996). Since the policies are wider in scope, they tend to cut across many projects, plans and programs. On the other hand, a program covers a number of projects, while a plan covers a number of programs. Therefore program, plan and policy-based environmental assessments are relevant to and have the capacity to cover more projects than the project-based assessments (Dijkstra T (2003).

When SEA is embedded in a country's legislature, it has the capacity to reduce costs associated with project assessment and costs, since some impacts assessed from earlier projects within that wider framework can be applied to a number of projects; this caters for a wider networking.   Similarly, through SEA, broad guidelines can be institutionalized to ascertain minimum standards of the environmental assessment. SEA, being wider in scope, is likely to cover a wider area, and can therefore have more capacity to handle impact assessments, which gives it more credibility (Dijkstra, 2003).

SEA also facilitates decision-making (especially once the analytical environmental assessment is incorporated), thereby reducing project delays and harmonizes the entire project cycle, so that the direction of any project is certain depending on its nature, location and scope. EA can give earlier guidelines, going by the experience from related or similar projects, on what EIA steps can be skipped, and which are mandatory. This saves time and resources, eases the screening and scoping stages, and makes it possible for the other stages of the EIA to be applied in time from the project conception through to the end.

This is likely to provide some social and economic gains to the community, and improve the general standard of living since the reduced cost of project implementation may benefit the users of the product or service from that project. Lastly, due to clear inscription of environmental assessment guidelines in national and regional records, wider participation in the process is likely, as the basis of each project will already be clear to a section of the stakeholders. This can act as a conflict-resolution strategy. However, these assessment guidelines can be availed in local languages by print and alternative media to facilitate wider participation, since in the developing countries, the EIA is either new, or non-existent (Lee and George, 2000). The SEA can therefore be a reasonable substitute for EIAs, a role they can perform even more effectively. In Botswana, for instance, EIA is not yet mandatory (GoB, 2003). Yet the number of new projects being developed are so many that unless their long-term repercussions are addressed early through alternative means such as SEA, there projects can only be time bombs.

Challenges in the EIA administration:
Lack of compliance by Government agencies:
The EIA process is nobly managed by a government parastatal, the authority. Whereas it is mandatory for all relevant projects to undergo EIA, the government institutions have never complied. This has been attributed to the following factors, as per personal observation and communication) (NEMA (2004, 2006, 2005, 2003a and 2003b),:
   a) Do not consider NEMA as superior to them to warrant NEMA involvement in their affairs- in an activity akin to supervision;
   b) NEMA has no machinery to ensure compliance by government agencies;
   c) The agencies are aware that the government has no capacity to take

disciplinary actions against a department of its own, e.g. will not close them down. Fear of closure by government for non-compliance is a major facilitator of compliance by non-government bodies.

d) Rate of non-compliance even by non-government bodies is still high, necessitation a close follow up. Thus the numbers overwhelm NEMA;

e) Political interference even in cases where due EIA process has been followed, thus intimidating the NEMA management and making it feel less likely to intervene in government projects;

f) Negative publicity by senior government ministers about NEMA as scaring away investors.

Up to April 2006, 1080 projects had undergone an environmental impact assessment. This is estimated to comprise about 75% of projects qualified for EIA. .

Corruption by proponent:
Many proponents try to bribe the EIA experts, sometimes successfully, with a view to excluding some adverse impacts from the study report. This is done by reducing the thoroughness of public participation through various mechanisms, including:
Selecting for the consultant his / her audience during the EIA process. This includes: who to talk to, interview, invite in public meetings, or include in group discussions. This has the capacity to exclude some concerns, or even some positive impacts of the project.

Possible Solution:

The consultant should stay beyond reproach, follow due process and conduct the work professionally.

Corruption by consultant:
While trying to make the proponent to pay more, some consultants arm twist the proponents by scaring them that the project has too many significant impacts, and stand no chance to be approved. They go ahead to tell them that if they paid extra, they would talk to NEMA so that the project goes through. The victims are largely illiterate and the very busy business people, who understand least of the existing legislation.

Ignorance of the public about their roles in EIA and EA.
Most of the time, when the stakeholder consultation and public participation process is on, many fear to give negative comments fearing that they would be incriminated.

Solution:
Many consultants have allayed these fears by first educating the public the importance of openness of all parties in the process. Holding the meetings on neutral grounds, as well as building the confidence of the publics, helps make them open.

Project gaps and dishonesty:
In some cases, the proponent has deliberately not given full, clear information on the project to the consultant, who then faces the public with incorrect or incomplete information. Based on this, the public cannot give proper feedback. In some cases, the proponent has not quite confirmed some details of the project. However, amidst consultant scramble for the projects, the

proponent only gives what they have. Further details of the project sometimes, even core aspects, come when it is too late.

To ensure an effective public participation the following could help:
a) Draft a project proposal: This should give all salient features of the proposed project in simple language, in local language as well. This will include maps and other visual aids (Lee and George (2000), Dijkstra (2003), and Canter (1977).
b) Form an acting project team: This will incorporate all obvious stakeholder representatives.
c) Identify all groups of stakeholders: Done by the acting project team. A combination of methods such as (i) those who volunteer themselves; (ii) those identified by the project team; (iii) those identified by other third parties e.g. key groups in the community; etc will be used.
d) Broaden the project team by incorporating newly identified stakeholders. Let each group of stakeholders be free to have any one of their own to represent them, even if it means rotational representation. But let it be known to the group about the importance of continuity in representation, as any break in communication and information flow may disorient the team.
e) Make a stakeholder consultation and CPP plan (done by the newly expanded project team / now called working project team (WPT)). This will involve a combination of interviews; questionnaires; rapid appraisal technique; organized sessions (with individuals, groups and entire community, i.e. the public), etc (Lee and George, 2000).

f) Make the program available to members in public places and strategic popular sites e.g. an area designated as an open door (a site where all information about the programme can be accessed by any interested party)

g) The program contents will be communicated to the public using different languages, fora and places. Verbal (door- to door), print (local newspapers, posters) and electronic media (Radio, Television, Email, internet website) will be used to communicate the message so that the schedules of CPP are properly publicized. Also, there will be community notice boards specifically for the project updates and information meant to make the community come forward to participate.

h) Short presentations like concerts, role plays etc in the electronic media (e.g. radio, television etc), will be made to attract the attention of the public towards the project, and inform them about its plan, and how it may affect them- both positively and negatively (Canter (1977); Dijkstra (2003). This is meant to attract attention, so that the public who are not aware that they may be affected may also come forward.

i) Once per day, during and for one week preceding any gathering, vehicles mounted with loudspeakers will be used to advertise the schedules.

j) Once the gathering is in place, participatory methods (e.g. use of small groups, role plays, community mapping etc) will be used to gather the concerns, fears, issues, and other information relevant to the project-now and in the future (Coates, 2003). Each group will be asked to select their secretary, who will put down all their points, and later one member of each group will be asked to present their report verbally to the audience. Group discussions will ensure gender sensitivity. Information from the Small groups will be augmented with extra information put in writing by individuals who may have points they may not want to raise in public (either due to their sensitivity, or due to personality /disposition of the individuals). Once verbal presentations are over, the written group reports will be submitted to the secretariat of the project (whose members will also be in different groups, rotating, with one of them as a facilitator)

Reporting in Environmental assessments
Report should be compiled using the small reports, and verbal presentations. This should be summarized in point form and its contents made public within a day. The report should be in as many languages as is feasible to facilitate sustainability and financial constraints. This repost should be available at the secretariat, and some copies should be availed free to the public.

CHAPTER 19: RESOURCES AND REFERENCES

1. ADB (1996) Economic Evaluation of Environmental Impacts: A Workbook, Asian Development Bank. Manila. (cited in www.environmental valuation and cost benefit news-academic study / journal article)
2. Afullo A (2004) Environmental and occupational health aspects of waste management in Okavango Delta, Maun, Botswana. A PhD Thesis, COU, Spain.
3. Afullo, A (2006) Nairobi Households' roles and responsibilities towards improved solid waste management . MSc Thesis, WEDC, Loughborough University, UK.
4. Alder, G. 1995. Tackling Poverty in Nairobi's Informal Settlements: Developing an institutional strategy. Environment and Urbanization, 7(2), 85-107.
5. Ali M, Cotton A and Westlake K (1999). Down to Earth: Solid waste disposal for low-income countries. WEDC and DFID, Loughborough, pp 67.
6. Ali, M (2003). Solid Waste Management: A WEDC Post-graduate module. WEDC Loughborough University, Leicestershire.
7. Ali, M (2003). Solid Waste Management: A WEDC Post-graduate module. WEDC Loughborough University, Leicestershire.
8. *APHRC (2002) Population and Health Dynamics in Nairobi's Informal Settlements,* African Population and Health Center (APHRC), Nairobi
9. Bartone, C., L. Leite, T. Triche and R. Schertenleib. 1991. Private sector participation in municipal solid waste service: Experiences in Latin America. Waste Management and Research, 9, 495-509.

10.   Bartone, C.R. (1991). "Keys to Success: Private Delivery of Municipal Solid Waste Services".Infrastructure Notes, Urban No. UE-3, Transportation, Water and Urban Development Department, World Bank.

11. Bartone, Carle R. (1997). *Strategies for Improving Urban Waste Management: Lessons From a Decade of World Bank Lending*. HazWaste World/Superfund XVII Conference, Washington, D.C. http://www.undp.org/ppp/library/files/barton0 1.pdf.

12.   Benard T, Mosepele K and Ramberg L (2003) (eds). Environmental monitoring of the tropical and Sub-tropical wetlands. Proceedings of a conference in Maun, Botswana (December 4-8, 2002). Harry oppenheimer Okavango Research centre (HOOR), University of Botswana. Okavango Report Series No. 1. ISBN 99912-949-0-2.

13.   Berg, B. 1989. Qualitative Research Methods for the Social Sciences. Toronto: Allyn and Bacon.

14. Bernstein, J. (2000). *A Toolkit for Social Assessment and Public Participation in Municipal Solid Waste Management*. Draft working paper prepared for the Urban Waste Management Thematic Group, The World Bank, Washington, D.C.

15.   Bhutan Thimphu (2000). Environmental Codes of Practice for Solid Waste Management in Urban Areas National Environment Commission

16. Bloom, David, N. Beede and E. David (1995). The Economics of Municipal Solid Waste. *The World Bank Research Observer* 10(2): 113-50.

17.   Botswana Daily News, 23 August 2000 Pg 3

18.   Botswana Daily News, April 23rd, 2002,

pg. 2

19. Botswana Daily News, pg.2, August 2nd, 2001

20. Botswana heritage (2000). Gaborone, Botswana pg 26-27.

21. Bubba, N. and D. Lamba. 1991. Urban management in Kenya. Environment and Urbanization, 3(1), 37-59.

22. Bushra, M. 1992. Recycling as a Necessary Part of Solid Waste Management: Some Case Studies. Unpublished research paper. ILO/UNDP Inter-regional Research Workshop. Nairobi: 6-10 April 1992.

23. Canter, L W (1996). Environmental Impact Assessment. McGraw Hill International Editions: Civil Engineering Series. 2nd edition. Singapore. ISBN 0-07-114103-0

24. CIEH (1998) Health And Safety: First Principles. The Chartered Institute of Environmental health (CIEH), Chadwick House Group Limited, London. ISBN 1 902423 00 7

25. CIEH (1999) Environmental Management Resource book for the Chartered Institute of Environmental health (CIEH), Chadwick House Group Ltd, London

26. CIEH (1999(b) Environmental Management: Training Plans for the Chartered Institute of Environmental health (CIEH), Chadwick House Group Ltd, London

27. City farmer (2003) In: http://www.cityfarmer.org/NairobiCompost.html

28. CLAUDIA S: 2000. Republic Of Namibia preliminary report of the water demand mgt study of the Namibian tourist facilities. Min of Agric, Water and Rural Development. Department of water affairs.

29. Cointreau, S., P. Gopalan and A. Coad (2000). Private Sector Participation in Municipal Solid Waste Management: Guidance Pack (5 volumes). SKAT, St. Gallen, Switzerland.

30. Cointreau-Levine, S. 1994. Private Sector Participation in Municipal Solid Waste Services in Developing Countries, Volume 1: The Formal Sector. UNDP/UNCHS/World Bank Urban Management Programme. http://www- Discussion Paper #13. Washington: The World Bank.

31. Crites R and Tchobanoglous G (1998). Small and decentralized wastewater Management systems. McGraw Hill international editions. McGraw Hill, Singapore. ISBN 0-07-289087-8

32. Deverill P, Bibby S, Wedgewood A, and Smout I (2002). Designing water supply and sanitation projects. WEDC, Loughborough.

33. Deverill P, Bibby S, Wedgewood A, and Smout I (2002). Designing water supply and sanitation projects. WEDC, Loughborough.

34. Dijkstra T (2003) Environmental Assessment. A WEDC Postgraduate module. WEDC Loughborough University, Leicestershire.

35. Durham N and Jackson M (2001) People and systems for water, sanitation and health: Monitoring water quality in the developing world. 27th WEDC conference, Lusaka, Zambia

36. Evans, H. 1992. A virtuous circle model of rural-urban development: evidence from a Kenyan small town and its hinterland. Journal of Development Studies, 28(4), 640-667.

37. Fisher A A, Freit J R, Laing J, Stoeckel J and Townsend J (1998). Handbook for family

planning operatrions research design. 2nd edition. New York population council

38. Fisher, J. 1993. The Road from Rio: Sustainable Development and the Nongovernmental Movement in the Third World. Westport, CT: Praeger.

39. Flintoff F (1976). Management of Solid wastes in developing countries. World Health organization. WHO regional Publications, South East Asia Series No.1.

40. Foundation for Sustainable Development in Africa. 1993. Composting for the Small Farmer: How to Make Fertilizer from Organic Waste. Nairobi: FSDA in collaboration with UNEP.

41. Freeman, D.B. 1991. A City of Farmers: Informal Urban Agriculture in the Open Spaces of Nairobi, Kenya. Montreal: McGill-Queen's University Press.

42. Gatheru, W. 1994. Community Small Scale Composting in Nairobi. Paper presented at the African Waste Forum, UNEP/UNCHS, Nairobi, November 1994.

43. Gathuru P (1990) Waste Recycling in Kitui-Pumwani, Wajibu Vol 5 No. 4 pages 20-21, 1990.

44. Gathuru P.K (1993). "The creation of an innovations programme for communities and micro entrepreneurs dealing with recovery and recycling of urban solid waste in Nairobi" A proposal-Undugu Society of Kenya, 1993.

45. Gathuru P.K. (1990) Waste Recycling in Kitui-Pumwani, Nairobi Wajibu Vol 5 No. 4 pages 20-21, 1990.

46. GoB (1992) Government of Botswana. Botswana National Water Master Plan. Final Report on phase 2. Volume 2.Ministry of Mineral Resources and water Affairs.

Department of water Affairs. Gaborone.

47.   GoB (1997) Government of the Republic of Botswana presentation to the 5th Session of the United Nations Commission on Sustainable Development. 1 April 1997

48.   GoB (1998(a)). Waste Management Act. Republic of Botswana, Gaborone. Government Printer.

49.   GoB (1998) Government of the Republic of Botswana presentation to the 5th and 6th Sessions of the United Nations Commission on Sustainable Development.. 9 February 1998.

50.   GoB (2000(a)). Botswana Environment Statistics. Central Statistics office, Gaborone, Botswana

51.GoK (1972). Public Health Act (Cap 242, Revised Edition). Republic of Kenya, Nairobi.

52.   GoK (1985). The Project for the Feasibility, Implementation and Management of the Collection, Transportation and Disposal of Urban Waste in Nairobi. Republic of Kenya. Nairobi.

53.   GoK (1994) The National Environment Management Plan (NEAP), Nairobi.

54.   GoK (1994a) Instituionalization of urban environment management, training and awareness creation. A project proposal to the Netherlands Government by ministry of Local government. Nairobi, Kenya.

55.   GoK (1994b) Institutionalization of urban environment management, training and awareness creation. Environment and urban development training project Phase II Nairobi, Kenya.

56.   GoK (1999).   The Environment Management and Co-ordination Act (EMCA), 1999. Nairobi

57. GoK (2001) Population and household census, 1999, Republic of Kenya. Ministry of planning and national development. Nairobi, Kenya.

58. GoK (2001a) Population and housing census-Vol 1. Ministry of Finance and national planning. Government press, Nairobi.

59. GoK (2001a) Population and housing census-Vol II.-Socioeconomic profile. Ministry of Finance and national planning. Government press, Nairobi.

60. GoK (2003) Kenya Demographic and health Survey report, Central bureau of statistics, Nairobi.

61. Haan, H.C., A. Coad and I. Lardinois (1998). *Municipal Solid Waste Management: Involving Micro- and Small Enterprises - Guidelines for Municipal Managers.* International Training Centre of the ILO, SKAT, WASTE, Turin, Italy.

62. Habitat *Privatization of Municipal Services in East Africa: A Governance Approach to Human Settlements Management.* Published by United Nations Centre for Human Settlements (Habitat), with support from the Ford Foundation, Office for Eastern Africa. Nairobi, Kenya. http://www.unchs.org/unchs/planning/privat/contents.htm

63. http.www.itcltd.com.waste disposal (1998)

64. http.www.unep.ch.elb.publications.econinst.Kenya (1998) Selection, Design and Implementation of Economic Instruments in the Solid Waste Management Sector in Kenya: The Case of Plastic Bags. UNEP

65. http://www.cityfarmer.org/NairobiCompost.html

66.   http://www.unep.org/PDF/Kenya_waste_m ngnt_sector/chapter3.pdf (2003)

67.   http:/www.ramsar.org/cop7_doc_20.5_e.ht m

68.   http:\\ www :Okavango Delta Peoples of Botswana.htm

69.   Huntsman-mapila P, Masundire H and Nyateka N (2003) Water Quality and Plankton in the Okavango Delta: Chapter 1: in Mosepele K (2003). HOORC, MAUN, Botswana. Unpublished.

70.   Ikiara, M.M., Karanja, A.M. and Davies, T.C. (2004). "Collection, Transportation and Disposal of UrbanSolid Waste in Nairobi", in Baud, I., Post, J. and Furedy, C. (eds.), *Solid Waste Management and Recycling: Actors, Partnerships and Policies in Hyderabad, India and Nairobi, Kenya*, Chapter 4, Kluwer Academic Publishers, Dordrecht, The Netherlands.

71.   Ikonya J N (1991) Refuse management in residential areas in Nairobi. A case study of Githurai and Umoja. An unpublished BA land economics dissertaton, UoN, Nairobi.

72.   *ILO* (2001) Employment Creation through Privatized Waste Collection, Kenya. - *Identification of Current SWM Practices and Scope for ILO Support in Nairobi Kenya*

73. Ince M and Howard G (1999) Integrated development for water supply and sanitation: Developing realistic drinking water quality standards. 25th WEDC conference, Addis Ababa, Ethiopia

74. IRC (1981) Small community water supplies- Technology of small water supply systems in developing countries. Technical paper series No 18 WHO, The Hague.

75. IRC (2002) Small community water supplies-technology, people and partnership. IRC Technical paper series No 40 WHO, The Hague

76. Itcltd (1998) In: http.www.itcltd.com.waste disposal (1998)

77. Iyer, Anjana 2001. *Community Participation in Waste Management: Experiences of a Pilot Project in Bangalore, India*. Urban Waste Expertise Program, the Netherlands. September. http://www.waste.nl/docpdf/CS_cp_bang.pdf

78. JICA (1998). "The Study on Solid Waste management in Nairobi City in the Republic of Kenya",Japan International Cooperation Agency.Nairobi

79. JICA, NCC, MOLG, Rep. Of Kenya, (1998) "The study on solid waste management in Nairobi C i t y i n t h e R e p u b l i c o f K e n y a : F i n a l R e p o r t V o l . 2 " , 1 9 9 8

80. Jones J A (1997). Global Hydrology: Processes, resources and environmental management. Pearson prentice Hall, ISBN 0 582 09861 0. Essex.

81. Kettel, B., B. Muirhead, J. Abwunza, G. Daly, J. Malombe, D. Morley, and P. Ngau. 1995. Urban Poverty and the Survival Strategies of the Urban Poor in Nairobi: Final Report. Toronto: Faculty of Environmental Studies, York University.

82. Kgathi D and Bolaane B (2001) Instruments for sustainable solid waste management in Botswana. Waste Manage Res 2001:19:342-353. ISWA. ISBN 0734-242X.

83. Kibwage J K. and Momanyi G M. (2003): The Role of Community Composting Groups in Nairobi cap 21: In: Canon E.N. Savala, Musa N. Omare and Paul L. Woomer (Eds).

Forum for Organic Resource Management and Agricultural Technologies (FORMAT), Nairobi.

84. Kibwage, J.K. 1996. Towards the Privatization of Household Solid Waste Management Services in the City of Nairobi. MPhil Thesis. Moi University. Eldoret, Kenya.

85. Kibwage, J.K. 2002. Integrating the Informal Recycling Sector into Solid Waste Management Planning in Nairobi City. PhD Thesis. Maseno University. Kisumu, Kenya.

86. Kim P (1998): Low Cost Solid Waste Incinerator: Demand Survey and Country Selection Report. Intermediate Technology Consultants Integrated Skills Limited. Mazingira Institute, the Faculty of Environmental Studies at York University, Toronto, Canada, the Department of Urban and Regional Planning at the University of Nairobi, the Undugu Society of Kenya, the Foundation for Sustainable Development in Africa, and the Uvumbuzi Club of Nairobi. in: http.www.itcltd.com.waste disposal

87. Komu M, Kimwandu N, Chriswa R.K and Kamau J.N (1993)"Report on the Task Force for Integration of Machuma Schools into Undugu Society of Kenya operation Oct.1993.

88. Kwach, O.H. and Antoine, P. (2000), "Mukuru Recycling Centre", UNCHS (Habitat) Report, July 2000.: In selection, Design and Implementation of Economic Instruments in the Solid Waste Management Sector in Kenya

89. Lamba, D. 1994. The forgotten half; environmental health in Nairobi's poverty areas. Environment and Urbanization, 6(1), 164-173.

90. Lapin, L. L. (1987). Statistics for modern business decisions. Harcourt Brace Jovanovich, Inc

91. Lardinois, Inge (1996). *Solid Waste Micro and Small Enterprises and Cooperatives in Latin America*. The Global Development Research Center. http://www.gdrc.org/uem/waste/swm-waste.html.

92. Lee N and George C (2000). Environmental Assessment in Developing and Transitional Countries. John Wiley and Sons, Manchester. ISBN 0-471-98557-0

93. Lee-Smith, D. 1990. Squatter landlords in Nairobi: a case study of Korogocho. In Amis, P. and P.Lloyd (eds.). Housing Africa's Urban Poor. New York: St. Martin's Press.

94. LEVINE UWP 1994 Micro-Enterprise Development for Primary Collection of Solid Waste

95. Liyai Khadaka (1988) Problems of SWM in the city of Nairobi. Unpublished MA Thesis, UoN, Nairobi.

96. Malombe, J (2001) "Employment Intensive and community based approaches to urban infrastructure development in informal settlements in Kenya", (Draft report for ILO)

97. Malombe, J. M. 1995. Approaches to Urban Poverty in Kenya: Non-governmental Organizations and Community Based Organizations. Draft Report. Ottawa: International Development Research Centre.

98. Marekia E.N and Gathuru P K (1990) "Waste Recycling in Nairobi" Undugu Society, Kenya.

99. Mary A-N & Negussie T (undated). The triad of poverty, environment and health in Nairobi informal settlements.

100. Matovu, G. (2000), *Upgrading urban low-income settlements in Africa: constraints, Potentials and policy options, Regional round able on upgrading low-income* settlements, Johannesburg, South Africa, Municipal development programme, Zimbabwe

101. Mbaiwa J (2003). The socio-economic sustainabiliyty of Tourism Development in the Okavango Delta, Botswana. Okavango Report Series No. 1. pp 83. HOORC, Maun, Botswana.

102. McCarthy T, Ellery W and Gieske A (1994). Possible groundwater pollution by sewage effluents at camps in the Okavango delta: Suggestions for its prevention". In: Botswana Notes and records, (26):129-138.

103. Medina, Martin (1997). *Informal Recycling and Collection of Solid Wastes in Developing countries: Issues and Opportunities*. United Nations University, Institute of Advanced Studies. http://www.gdrc.org/uem/waste/swm-ias.pdf.

104. Memon, P. 1982. The growth of low-income settlements: Planning responses in the peri urban zone of Nairobi. Third World Planning Review, 4(2), 145-158.

105. Mitlin, D. 1990. Human Settlements and Sustainable Development. Nairobi: United Nations Centre for Human Settlements (Habitat).

106. Mitullah Winnie 2003 Urban Slums Reports: The case of Nairobi, Kenya IN: understanding slums: Case Studies for the Global Report on Human Settlements UNHCR / Habitat, Nairobi.

107. Mmegi, 22-28 February 2002, pg.13

108. Mmegi, 22-28 February 2002, pg.13

109. Mosepele K (2003) (ed). Aquarap II. Rapid

assessment of the Aquatic ecosystems of the Okavango Delta, Botswana, Jan 31 – Feb 17, 2003. preliminary Report. Conservation International, Botswana.

110. Moser, C. 1989. Community Participation in Urban Projects in the Third World. Toronto: Permagon.

111. Mugenda O and Mugenda A (1999). Research methods: A quantitative and qualitative approach. ACTS Press, Nairobi.

112. Mulaku, G C, and Siriba, D N.(undated) Identification of Alternative Solid Waste Disposal Sites in Nairobi, Kenya: The GIS Approach In: Journal of Civil Engineering Research and Practice

113. Mulei, A. and Bokea, C. (eds.). (1999). Micro and Small Enterprises in Kenya: Agenda for Improving the Policy Environment. The International Center for Economic Growth (ICEG), Nairobi.

114. Munro, B. 1992. A new approach to youth activities and environmental clean-up: The Mathare Youth Sports Association in Kenya. Environment and Urbanization, 4(2), 207-209.

115. Munyakho, D. 1989. Kenya: The parking boys of Nairobi. In D. de Silva (ed.), Against All Odds: Breaking the Poverty Trap. London: Camelot Press.

116. Mwaura, P.M. 1991. An Assessment of the Management of Garbage Collection and Disposal in Nairobi. Nairobi: Department of Urban and Regional Planning, University of Nairobi.

117. Nairobi District Development plan (2004-2008). Ministry of planning and national development. Government printers, Nairobi.

118. NCC (2001). "Policy on Private Sector Involvement in Solid Waste Management",

Department of Environment, Nairobi City Council.

119. NCSA (undated) National Conservation Strategy Agency (NCSA). Ministry of Lands, Housing and Environment brochure.

120. Ndungu A (2006) Hydrogeological survey report. Kipsitet, Kericho district.

121. NEMA (2003a) Environmental Impact Assessment (EIA) brochure, Nairobi, Kenya

122. NEMA (2003b) Environmental Audit (EA) brochure, Nairobi, Kenya

123. NEMA (2004) Personal communication with a provincial NEMA coordinator, Kisumu, Kenya

124. NEMA (2005) Personal communication with a national EIA/EA coordinator, Nairobi, Kenya

125. NEMA (2005). National Environment Management Authority news Margazine. A quarterly publication for NEMA. Nairobi, kenya

126. NEMA (2006) Personal communication with a provincial NEMA coordinator, Kakamega, Kenya

127. NWDC (2000). North West District Council. Preliminary feasibility study report of a proposed new landfill site for Maun Village. Report prepared by Group Consult, Francistown.

128. NWDC (2002). Investigations of bad smell and structural failure of Maun Sewerage Pump Stations. Draft Final Investigation Report. May 2002. By group Consult Botswana, Francistown.

129. Odegi-Awuondo, C. 1994. Economics of garbage collection: A survival strategy for Nairobi's Urban Poor, in C. Odegi-Awuondo et al. (eds.), Masters of Survival. Nairobi: Basic Books Kenya, 45-62.

130. Ondiege, P. and P. Syagga. 1990. Nairobi City socio-economic profile, in J. Odada and J. Otieno, (eds.), Socio-Economic Profiles, Nairobi. UNICEF/Government of Kenya.

131. Patton, M.Q.(1990). Qualitative evaluation and research methods. SAGE Publications. Newbury Park London New Delhi.

132. Peters, K. (1998). Community-based waste management for environmental management and income generation in low-income areas: A case study of Nairobi, Kenya. City Farmer. Toronto, Canada.

133. PriceWaterHouseCoopers (2003) Water supply and Sanitation, Building Kenya Together, Conference on private Sector participation in Kenya's Infrastructure, 15 May 2003

134. RAMSAR (1999) The ramsar Convention on wetlands: Ramsar COP7DOC.20.5: International Cooperation for the management of the Okavango Basin and Delta.UN.

135. Rubinda Mayugi & Kivasi, ILO 1994

136. Ruto D K B (1988). An investigation of solid waste collection and disposal systems. A case of Nairobi City commission. An Unpublished MSc Thesis, University of Nairobi. Nairobi, Kenya.

137. SADC, IUCN and SARDC (1994). State of the Environment in Southern Africa. Southern African Development Community (SADC), Southern African Research and Documentation centre (SARDC), and the World Conservation Union (IUCN).

138. Salant, P. and D. A. Dillman (1994). How to conduct your own survey. John Wiley & Sons, Inc.

139. Satterthwaite, D. 1993. The impact on health of urban environments. Environment and Urbanization, 5(2), 87-111.

140. Schmink, M. 1989. Community management of waste recycling in Mexico: The SIRDO. In A. Leonard (ed.), Seeds: Supporting Women's Work in the Third World. New York: The Feminist Press.

141. Schübeler Peter, Wehrle Karl and Christen Jürg, (1996). Conceptual Framework for Municipal Solid Waste Management In Low-Income Countries. Working Paper No. 9 UMP, Nairobi.

142. Serumola O and Mbongwe B (?) Chemical Substances Management in Botswana. An unpublished Conference paper.

143. Serumola O and Mbongwe B (2002) Chemical Substances Management in Botswana. Downloaded from http://irptc.unep.ch/POPs_nc/proceedings/Lusaka/SERUMOLA.html

144. Sethuraman, S.V. 1981. The Urban Informal Sector in Developing Countries: Employment, Poverty, and Environment. Geneva: International Labour Office.

145. Smith M (2000). Water and Environmental health. A WEDC Post-graduate module. WEDC, Loughborough University pp 4.6-4.7 and 5.1-5.25.

146. Smoke, P. 1993. Institutionalizing decentralized project planning under Kenya's programme for rural trade and production centres. Third World Planning Review, 15(3), 221-247.

147. Smoke, P. 1994. Local Government Finance in Developing Countries: The Case of Kenya. Nairobi: Oxford University Press.

148. ST (1996) Somarelang Tikologo Strategiv plan. Gaborone, Botswana. Somarelang Tikologo.

149. ST (2003) Somarelang Tikologo. Spring / Autumn issue.Gaborone

150. Stren, R. (ed.). 1994. Urban Research in the Developing World, Volume 2: Africa. Toronto: Centre for Urban and Community Studies.

151. Stren, R. 1991. Old wine in new bottles? An overview of Africa's urban problems and the "urban management" approach to dealing with them. Environment and Urbanization, 3(1), 9-22.

152. Sue Coates (2003) Community and management. A WEDC Postgraduate module. WEDC, Loughborough University.

153. Syagga, P. 1992. Problems of Solid Waste Management in Urban Residential Areas in Kenya. In The Proceedings of African Research Network for Urban Management (ARNUM) Workshop: Urban Management in Kenya, Joyce Malombe (ed.). University of Nairobi, August 20, 1992.

154. Syagga, P. 1993. Background Paper on Waste Management in Central and Eastern African Region. Unpublished research paper. Nairobi: University of Nairobi, Department of Land Development.

155. Tchobanoglous G, Theisen H and Vigil S. (1993). Integrated Solid Waste Management" Engineering principles and management issues. Mc-Graw Hill International Editions. Mc-Graw Hill, New York.

156. Tenambergen, E. (1997) Solid waste management in Nairobi City: knowledge and

attitudes. Journal of Environmental Health; 12/1/1997.

157. Thurgood Maggie (undated) Solid waste Landfills in Middle and lower income countries: A technical guide to planning, design and operation

158. Thurgood, M., ed. (1999). *Decision-maker's Guide to Solid Waste Landfills: Summary*. Transport, Water and Urban Development Department, The World Bank, Washington, D.C.

159. UCCAS, ILO 1999private Sector Participation in Municipal Solid Waste Services in Developing Countries Vol. 1: Formal Sector,

160. UNEP (1998) In: http.www.unep.ch.elb.publications.econinst. Kenya Selection, Design and Implementation of Economic Instruments in the Solid Waste Management Sector in Kenya: The Case of Plastic Bags. United Nations Environment Program, Nairobi, Kenya

161. UNEP (2003) In: http://www.unep.org/PDF/Kenya_waste_mng nt_sector/chapter3.pdf (2003)

162. UNEP (2003) UNEP in 2002. Environment for development. UNEP, Nairobi

163. UNEP (2004). Selection, Design and Implementation of Economic Instruments in the Solid Waste Management Sector in Kenya. The Case of Plastic Bags. UNEP, Nairobi.

164. UNEP and ACTS (2001) The making of a framework Environmental law in Kenya. ACTS press, Nairobi.

165. UNHCR (1989). Solid Waste Management in Low Income Housing Projects: The Scope for Community Participation. United Nations-Habitat Programme, Nairobi, Kenya.

166. UNHCS. *City Garbage Recyclers: Kenya.* Best Practices Database. http://www.bestpractices.org/cgi-bin/bp98.cgi?cmd=detail&id=5562. (Accessed July 2, 2001) Together Foundation/UNHCS

167. Wachira J.W. (1994) Undugu Society of Kenya Dandora Rescue Centre "Waste paper group project report April, 1994.

168. WEDC (2004) Integrated Water Resources Management (IWRM): A WEDC Postgraduate module. WEDC, Loughborough University, UK.

169. WEDC (undated) Community and management: Additional resources. Loughborough University, UK.

170. World Bank, Swiss agency for Development & cooperation, WHO and Swiss centre for Development cooperation in technology and management. St gallen, Switzerland.

171. WSP (2005) Understanding Small Scale Providers of Sanitation Services: A case Study of Kibera, Field note, Water and Sanitation Programme, Nairobi Kenya.

Made in the USA
San Bernardino, CA
14 March 2014